Sense of Serving

FSC
www.fsc.org

Sense of Serving
Reconsidering the Role of Universities Now

Meindert Flikkema (ed.)

VU University Press, Amsterdam

What's your opinion about a new role of universities in contemporary society?
Please share it with us at www.senseofserving.com

VU University Press
De Boelelaan 1105
1081 HV Amsterdam
The Netherlands

info@vu-uitgeverij.nl
www.vuuniversitypress.com

Design jacket: Studio Marion Rosendahl, Vreeland
Type setting: Titus Schulz, Arnhem

ISBN 978 90 8659 731 4
NUR 840

Preface

JAAP WINTER

The contributions in this book express deep concerns about our current university system. They mention the massification of education, the under-appreciation of education compared to research, a focus on managerial output, the ever-increasing competition for research funding, a growing disconnection between research and education, excessive external accountability and the resulting increased levels of bureaucracy. These developments have generated an environment where teaching and research have come to resemble a production process where there is little to no scope for student formation beyond the mere acquisition of knowledge and where teachers find it more and more difficult to provide inspiration to their students and to remain inspired themselves.

These concerns are not new; they have been voiced by many others. What is so special about this particular book, with its focus on education, is that the authors move beyond complaining about the current system and offer concrete and specific examples of how things could be done differently: better and more inspiring. Although the challenge may seem daunting, the system overwhelming and its core elements beyond our control, in reality there is much that we can do. We *are*, in a way, the system. It helps if we do not perceive what we are facing as a system outside ourselves, but learn to see it as a set of human practices. We are in charge of these practices ourselves. We can make our educational practices more human again, giving the human effort of teaching and learning a central position. Each of us can make some change somewhere:

* Professor Jaap Winter is President of the Executive Board at Vrije Universiteit Amsterdam.

sometimes by taking small steps at a time, as Frank van der Duyn Schouten explains in his contribution. In addition, we should challenge the assumptions underlying the current set of practices and the values they represent. As Steven ten Have writes in his contribution, technical changes (how we do things) are not going to assist us when the nature of the underlying challenge is adaptive (why do we do these things?). In fact, we need to adapt ourselves – and that may well be the hardest part. Change will therefore take time, but I have no doubt that we can make fundamental changes together, even within an external framework for higher education that we cannot change on our own.

At Vrije Universiteit Amsterdam we should take courage from our name, as Halleh Ghorashi so rightly suggests in her contribution. We are free to experiment, to try, to sometimes simply ignore what the system seems to require, to be different. In the process, we are also free to make mistakes and learn from them, all in order to become the university we want to be. Feeling free to do so, keeping in mind what ultimate purpose we are trying to achieve, will unlock the passion that the authors of this book so clearly display. That ultimate purpose as we feel it at Vrije Universiteit Amsterdam is to contribute to a better world, a world in which justice, common humanity and a responsibility for each other and the world are central. We seek to do this through our education, our research, through sharing knowledge with society and serving society; in doing this, we value attention to different views of life and finding meaning. We shall use these words to describe the purpose of Vrije Universiteit Amsterdam in our new constitution: a purpose that will encourage all of us to maintain a sense of serving in everything we do.

I thank Meindert Flikkema, the initiator and editor of this book, for his initiative and drive to find ways for us to become that university. And I thank the authors for their contributions. They will change our thinking about what can be done so that we may start to change what we do.

Contents

From the Editor

MEINDERT FLIKKEMA

Great teachers care

Some time ago, in the spring of 2014, I was interviewed about the Faculty of Economics Education Prize that I had been awarded. When the interviewer asked me 'What is it that made you win this prize and that makes you "best" teacher?' I found the question rather hard to answer. I still find it hard to answer, to be honest, because I believe that variety is the spice of life – and educational life, too. I continue to discover new educational challenges, and this is partly the reason why I always feel slightly embarrassed by the use of the word 'best' in the phrase 'best teacher'. In any case, the interview question made me a bit restless. I found myself moving backwards and forwards in my chair, but I nevertheless decided to wait for as long as it would take for a satisfactory answer to pop up. And it did: 'Great teachers care!' is what I replied, without directly referring to me personally. My immediate thoughts went to my mentor and colleague Peter Tack M.Sc. and to my colleague Dr. Donna Driver-Zwartkruis, both the epitome of caring teachers. 'Great teachers care,' I said to the interviewer, 'now there's a topic that would deserve a good book.' And the next day I decided to write that book myself. I felt inspired by my own unexpected thought, the unthought known. Ever since that moment, 'Great teachers care' has been my motto. And it may even prove to be the decisive 'winner' in assessing the quality of 'schoolteachers': educators, instructors, lecturers and professors. Great teachers teach

* Dr. Meindert Flikkema is associate professor in the Faculty of Economics and Business Administration at Vrije Universiteit Amsterdam.

not for their own glory, but to serve others, namely those who have been entrusted to their care: their pupils and their students.

You are a great teacher, but …

During a job performance review to which I was invited after I had talked to the journalist, I was again complimented on my educational skills, but I also detected a few critical words here and there. I was admonished, and not for the first time, I might add, for my shortcomings in terms of research output and my lack of societal profile. The first thought that crossed my mind upon hearing this was 'But if that is the case, then probably none of us have enough societal profile' and 'It's not someone's societal profile that matters, but their meaning to society – or "impact" if you prefer this term.' Here, we mustn't forget that 'impact' in no way refers to monetary rewards. If we fail to mind our ways, society will be degraded to a mere cash dispenser from which research funds are simply 'withdrawn': not for fun, of course, but to survive in today's rat race in Academia. This rat race and its effects can be seen everywhere we look, and the race continues to produce highly competent and socially highly meaningful 'losers'. These 'losers' are people who often refuse to accept systems in which ideals are sacrificed for the sake of the wish to 'measure' things and to be 'in control'. These so-called losers are my heroes. Attempts have been made to silence their ideals, to corrode their character, through systems driven by perverse incentives: a despicable and unloving endeavour. We have to try and turn the tide, and we have to do this lovingly.

Creating the right conditions for learning

Could it be that we, as residents of Academia, are becoming a little too fixated on impact factors and the production of papers without considering their meaning for society? We *can* make an impact and have meaning in the classroom, at least if we don't engage too much in talking about what we know and what we think, or what we think we know. This may sound a bit facetious, and I am not saying that knowing everything can't make a person happy, but what I mean to say is related to my conviction that teachers should create the right *conditions for learning* whenever and wherever they can. The one thing they shouldn't do is overwhelm their students in one-way communication. At our official Opening of the Academic Year in 2014, I said that it would be a good idea to turn each lecture into a class. This is not because I'm a lazy person, which I am not, but because people learn much more by experiencing things rather than by listening to someone talking about things, especially if the listening has to be done for an impressive number of hours. And even when this cannot be avoided, people prefer to listen to narratives filled with inspirational and rich analogies, as Steven ten Have so beautifully explains elsewhere in this publication. These narratives are the type of stories that resonate with people and that they will remember. The rest will evaporate quickly.

Upon leaving the room where I had been discussing my performance, I remarked 'You will be pleasantly surprised!' With this promise in mind I started working on the current publication, *Sense of Serving*, and I included a chapter entitled *Great Teachers Care*. Little did I know at the time that 'caring' would prove to be one of the most important adjectives and verb forms in the entire collection of essays. My quest started with finding inspirational colleagues who after reading my essay 'Sense of Serving' would be prepared to join me as my co-authors. Finding them proved to be easy: none of them objected to my goals and none of them declined my invitation. The fire had been lit and instantly started to burn brightly.

Sense of Serving

The word 'serving' is not synonymous with 'servicing'. This means that the essays brought together in *Sense of Serving* make no plea for turning the University into a professional service organization or an even more professional one, for that matter. Readers will not be presented with a total quality management approach, nor will they find an appeal for the large-scale or small-scale appointment of education professors. What they will find is a call to determine *purpose*, to define what we are jointly working for, what we are working on together, what we fight for, what we pray for, what we acknowledge rather than reject, and what we respectfully refer to. Readers will be presented with a guiding principle that offers inspiration and transformation, that connects and offers forgiveness, that is sustainable and that will last. This principle is no single mark on the horizon, but represents the entire horizon: an inviting perspective that keeps changing its nature thanks to developments in society but that nevertheless stays true to its core meaning. As far as we are concerned, a strong 'Sense of Serving Society' could act as a catalyst for the transition that Science needs to make. This requires servant leadership at all levels of the University. It also requires the courage to combat or ignore institutions that attempt to undermine our purpose. Finally, it requires the courage to reject megalomania and other forms of deviant behaviour. What really matters are the meaning and the impact of the community of learners as a whole. The potential of this community is far greater than that of single individuals.

Sense of Serving, the book now lying in your hands, was written by nineteen passionate and highly driven teachers, academics, and students I had the honour of knowing when I taught the Managing Service Innovation course in their Master's programmes and when I taught post-graduate students in the Executive Master Programme Management Consulting. They were invited to write an essay about their vision of education as it should be offered by future-proof universities. I then continued to work with the winners of this essay contest in a way described elsewhere in this publication, acting as a *more capable peer*. The experience was memorable, educational and inspiring. These students' perceptions and the topicality of their experiences gave me much appreciated new insights, and their gratitude has motivated me to continue my work.

This book is offered as a gift. I present it to all my academic colleagues at Vrije Universiteit Amsterdam and VU-mc in Amsterdam, to all managers operating in Dutch Academic Education and to many, many students. I do this not to impose myself on them, but to ask them in all humility to read our essays carefully and attentively – and to talk about them with others. If this can be done, we will be able to liberate ourselves from any institution that hinders the rise and further development of universities that are truly *free*, universities that put student development centre stage in their words as well as their actions. In this way, they will carry much greater meaning for society, and their research will have a far greater societal impact than ever before.

A word of thanks

I would like to thank all those who have contributed to the publication of this book: my co-authors Sjoerd Martens, Nishand

Sardjoe, Tom Zegers M.Sc., Lita Napitupulu, Anne Laurette Leijser, Osman Aksoycan, Laurens Walinga, Loes Hafkamp, Prof. Halleh Ghorashi, Prof. Tineke Abma, Dr. Sylvia Van de Bunt, Prof. Frank Van der Duijn Schouten, Prof. Jos Beishuizen, Prof. James Kennedy, Prof. Steven ten Have, Prof. Fons Trompenaars, Dr. Elco van Burg, Prof. Mario van Vliet and Prof. Johan Wempe. Thank you all for joining me in a shared attempt to light a comfortable fire and for your highly motivational contributions. I also thank Laetis Kuipers-Alting M.A. for her excellent textual 'make-overs', as Frank van der Duijn Schouten so aptly phrased it, and for checking grammar and style. Thanks also go to Dr. Jan Oegema, editor at VU University Press, whose dedication and awareness of the core essence of our message have made it possible for us to follow this singular publication strategy. Finally, I would like to thank Prof. Jaap Winter, President of the Executive Board at Vrije Universiteit Amsterdam, not only for his fitting and eloquent preface to this book but also for his inspirational and binding role in the continued development of our 135-year-old university.

Amersfoort, March 2016

Sense of Serving

MEINDERT FLIKKEMA

The current situation at many Dutch universities is one of growing dissatisfaction. Internal as well as external parties are voicing their discontent increasingly loudly, with the recent student occupations of the University of Amsterdam's Bungehuis and Maagdenhuis[1] and the subsequent eviction of protesters by the police illustrating how bad things have become. What lies at the heart of today's widespread discontentedness is the fact that universities have become estranged from their roots: the *universitas*. This was the concept that reflected the founding ideas of the first European universities, established as early as the twelfth century, and that placed a primary focus on education. Over time, however, this focus has shifted towards research, and particularly towards measurable research output. What counts heavily today is the number of academic publications that researchers produce and their impact as determined by academic peers world-wide. The related competition for resources, tenure and promotion is strong and a severe threat to top-management claims that universities are a community (of learners).

The dominance of research has become institutionalised. It has led – and continues to lead – to strategic behaviour, to the dehumanisation of work and the working environment, and also to the erosion of education quality. Although many people, fuelled by their dissatisfaction, are striving to find what they call the right balance between research and education, the use of the term 'balance' would not seem to be quite appropriate here. The key word is

1 For more information, see https://en.wikipedia.org/wiki/Bungehuis_and_Maagdenhuis_occupations.

'reconciliation'. The reconciliation of research and education can reverse the process of alienation currently witnessed. Admittedly, some degree of change can be discerned in our universities, but developments have so far remained rather incremental. After all, what was won in the past is not given up lightly, and a certain price must be paid for this. We may conclude that Science has not yet made the transition it so badly needs, as a result of which the moral aspirations, ideals and core values of universities have become increasingly undermined.

What is needed to turn the tide is a velvet revolution that will breed new collective ideas and convictions. This type of revolution will be launched from within the academic world itself, not to be broken by anyone. Love and devotion cannot be broken. Universities that are built to last and stand the test of time will make sure that education re-occupies centre stage to form the basis of their work. What truly matters in this process is not expressed in terms of the number of contact hours, but in terms of their contact quality. We need to reach out and promote true human contact and interaction between junior and senior academics. In this way, the gap between research and education will disappear. As a matter of course, experts from all over the world will be offered a chance to share their deep knowledge, for instance via Massive Open Online Courses. Let's call this 'science teaching'. It means that other people's expertise will be redirected towards creating conditions for experiential learning. Through their work, these experts will make sure that students literally experience how the vast potential of knowledge, something I prefer to conceptualise as 'supposed knowing', can be realised. This will generate wisdom and gratitude, but it will also lead to a sense of doubt. Precisely this doubt is intended to galvanise students into productive action rather than seduce them to form instant opinions. The core elements of this new way of learning include a wish to serve, and they demonstrate courtesy, truthfulness, dignity, and diligent inactivity. The active application of these virtues and the ability to appreciate the nourishment they offer will not only support and enrich the process of truth seeking, but it will also boost confidence and enhance social cohesion – inside as well as outside the academic community.

A university that is fit for the future will be transformed from a mere place where people simply come and go into the smithy that it needs to be, where much is required and expected from students, and quite rightly so. Learning from teachers and instructors, the more capable peers, will take place in a safe environment in which students may learn in ways that are 'free from church and state influence' and in which our academic ideals are actively expanded by the notion that the learning process should also remain 'free from egos who claim absolute truth' based on a strong track record in publishing from a reductionistic perspective.

The resulting new ecosystem that will emerge in future-proof universities will make our society more resilient and a better place to live, and it will never cease to feed Academia with questions that really matter and that carry true meaning for those who share this world. The new ecosystem will act as an (in)visible hand guiding societal developments, thus forming the best possible governance structure for society. A university that can stand the test of time does not *consider* such an ecosystem, but it nourishes and shapes it: it partly *is* this system, the incubator of a new world order.

A university that is future proof is not only receptive of the potential of new generations and capable of distinguishing trends and undertows in these groups, but it is also naturally inclined to respect what has to remain unchanged. This requires a clear sense of heritage and historical perspective. The current lack of *Bildung,* however, that many of today's students demonstrate cannot be solved exclusively via individual minor programmes. Instead, it should be promoted anywhere and everywhere, reflected in carefully designed teaching methods and styles that make the best possible use of modern technologies and set-ups in which the physical presence of mentors and fellow students is actively encouraged and cherished. The form in which content is offered has a significantly greater impact on the development of students than content alone. It is predominantly form that determines formation.

A future-proof university remains in constant connection with its roots and honours the *universitas* like a father and a mother. It is concerned with slow science, it shapes today's *homines universales,* and it returns to them the control they deserve to have over their learning processes. It shines a different light on society whenever this is needed, and it is not led by institutions that threaten its identity. It has liberated itself from the socially harmful and crippling conviction that 'only numbers tell the tale', and it puts codified knowledge in the right perspective. It preaches the principle of 'experiencing through practising', and it helps to develop talent in younger as well as older students by knowing them and by taking good care of them.

Great teachers care! This notion is by no means soft and sentimental; it is of existential importance and the very reason why a future-proof university's training courses for secondary school teachers should in fact become its flagship programmes. Such a university will move forward with determination and dignity. It will stop trying to carve isolated and meaningless pieces from the puzzles presented by a reality that is multi-interpretable. Its collective impact factor will be determined by the devotion of its staff towards societal development. We have already formulated a meaningful description of this idea: 'foundations for a caring society', a slogan recently published on signs adorning the front of the VU's main building. This is a slogan that we must never replace! It sets an example for our students to follow.

Today's exciting changes ask all of our university's staff members to develop a greater Sense of Serving: being more concerned with and empathic towards each other, students, and society as a whole. They also require managers to recognise that the realisation of more Sense of Serving in universities should in fact be their sense-of-self, their vocation rather than idle talk and mere theory. They need to be aware that they should primarily serve the development of their staff and also act as shepherds who look after their flocks. These conceptions will also form a source of inspiration in our work for students. The care and attention thus received will leave an indelible impression on these students. As academically educated individuals and citizens, they will accept their responsibility in combating the effects of our current meritocracy and in halting the polarisation it causes. They will do this with determination and compassion.

Sense of Serving means taking others into consideration and caring for them, as opposed to following an ego-centric way of working in which colleagues are seen as competitors. It means working together, within and across disciplines, for the further good of society. It offers us an opportunity to re-introduce passion and devotion into our work, and it enhances our sense of community. Sense of Serving implies reducing our speed and increasing our impact on society.

It only takes a few little pushes to open up a window of opportunity for universities to become liberated and free. These little pushes will spark a process of autopoiesis, of self-organisation, with Sense of Serving as a beacon guiding the academic ship towards the great and wonderful unknown rather than towards the emptiness that characterises the obsessive hunt for positions at the top of the academic pyramid. A university with promising potential for the future will rid itself of all the systems and symbols associated with such a fruitless quest. It will halt today's collective rush for gold in Academia. Craftsmanship is capable of organising itself and true mastery follows naturally.

Sense of Serving guides and acts: it connects, it builds, it transforms, and it adds value. Let us make a start by liberating ourselves from everything that threatens our academic freedom to 'improve the human condition' (Arendt, 1958 and 2013) This can be done as of today by placing the development of our students firmly at the centre of our ambitions, in our words as well as in our actions.

There is one particular opportunity that is waiting to be seized, namely the opportunity to re-adopt a pioneering role in the academic world and to present ourselves, justly and with a modest sense of pride, as the type of university that lives up to its name of *Vrije* Universiteit: a Free University laying foundations for a caring society that are solid as a rock.

References

Arendt, H. (2013). *The human condition*. University of Chicago Press.

The Schools of Athens: Educating Students to Develop Knowledge for Life

JOS BEISHUIZEN

Summary

What are universities for? This question echoes in many debates on the task and position of universities in a world in which profitability appears to be the one and only purpose. Chris Brink argues that universities should produce 'knowledge for life'. Meindert Flikkema advocates a sense of serving as a fundamental attitude of teachers and students alike. In this paper, I shall discuss three metaphors, each of which refers to an educational strategy for universities: the knowledge acquisition metaphor, the participation metaphor and the knowledge creation metaphor. I shall argue that preparing and enabling students in communities of learners to create knowledge for their own life and for the public good makes them autonomous, competent and caring academic citizens who will be able to withstand the profitability push.

Introduction

In his Vice-Chancellor's lecture held in 2007, Chris Brink (2007) raised the following question: what are universities for? Using Raphael's famous School of Athens fresco, with Plato and Aristotle situated in the middle of the scene, he explored two possible options: developing academic knowledge (following Plato) and educating academic professionals (following Aristotle). Brink concluded that both objectives are relevant. Universities should produce 'knowl-

* Jos Beishuizen is professor emeritus of Higher Education in the Department of Educational Sciences – Vrije Universiteit Amsterdam.

edge for life' (Brink, 2007, p. 9). In my contribution, I shall take Brink's conclusion as a point of departure, and my focus lies on educational strategies that universities should deploy to educate students so that they may acquire knowledge for life.

Knowledge acquisition and professional participation

The traditional view of the process of knowledge acquisition can be characterized by the acquisition metaphor (Sfard, 1998). This metaphor refers to the traditional view on learning as filling a container, the learner's memory. This container may be considered empty or filled with preconceptions or pre-existing knowledge. One of the most important findings in the domain of educational psychology was formulated by Ausubel (1968): '*The most important single factor influencing learning is what the learner already knows. Ascertain this and teach him accordingly*' (p. 68). This traditional view on learning is complemented by a view on the purpose of education as connecting individual students to the canon of academic knowledge (Joseph, 2011). This view on the process and purpose of knowledge acquisition is congruent with the first objective of universities, as distinguished by Brink (2007): acquiring and advancing academic knowledge. As Brink (2007) argues, this objective is also characteristic of liberal education, currently visible in liberal arts colleges or university colleges. Nussbaum (2012) made a strong plea in favour of liberal education as opposed to education with a predominant focus on the economic value of universities as schools where students are supposed to acquire profitable skills. This controversy has been the subject of many recent academic debates on the future of universities.

Flikkema's thought-provoking essay on the *Sense of Serving* (this volume) provides a valuable contribution to this debate. From his discourse, I shall extract five statements to which I will return later:

1 Education and research should be connected, reconciled.
2 Great teachers care.
3 Teachers, students, researchers and other members of the academic community all serve various functions.
4 A sense of doubt protects students from the tendency to 'form instant opinions'.

5 The academic community has a 'clear sense of heritage and historical perspective'.

Apart from the knowledge acquisition metaphor, there is a second metaphor (Sfard, 1998) that has inspired teachers and designers of academic curricula: the participation metaphor. In line with this metaphor, learning is defined as the gradual participation of students to their future professional world (Volman, 2006). This metaphor is related to what Brink (2007) mentions as the second purpose of universities: preparing students for life as academic professionals. Medical students prepare for their professional life as a doctor by serving as an intern in a hospital. Law students prepare for their professional life as a lawyer by serving as an intern in a law firm. In this way, students not only acquire disciplinary knowledge and skills, but they also participate in a peripheral way (Lave & Wenger, 1991) in their future professional world. They adopt the language, habits and culture of their future profession, and they start to develop a professional identity. Universities serve as a bridge between the academic world and the professional world.

The message seems straightforward: universities with their task to advance knowledge for life should follow both the acquisition metaphor and the participation metaphor. The acquisition metaphor is predominant in Bachelor's programmes, which are aimed at the acquisition of academic knowledge. Gradually, the participation metaphor gains more and more weight until it becomes the prevailing approach in Master's programmes, aimed at preparing students for an academic profession. This is what most universities have been doing for a long time. Unfortunately, today this mix of knowledge acquisition and professional participation does not appear to be able withstand the drive towards profitability.

Knowledge creation in communities of learners as a third alternative

There is a third metaphor: the knowledge creation metaphor (Paavola, Lipponen & Hakkarainen, 2004). According to this metaphor, students are challenged to engage in a process of constructing understanding (Joseph, 2011). This process is not necessarily and not always an individual process. Instead, students can collaborate with other students and with teachers or researchers in communities of learners (Brown & Campione, 1996; Beishuizen, 2004). In these communities, participants have various roles and operate at various levels of expertise. In Flikkema's words, *teachers, students, researchers and other members of the academic community all serve various functions*. Participants in a community of learners take each other seriously as partners in the creation of knowledge. In that sense, they care for each other. As Flikkema underlined, *great teachers care* for their students. In a similar vein, great students care for each other and for their teachers. In communities of learners, participants create knowledge by conducting research which helps them to understand their discipline (Healey, 2005).

Participants may also work on projects in the professional world of their discipline and, in this way, participate on a peripheral basis in the world of their future profession (Lave & Wenger, 1991). This characteristic of communities of learners supports the *reconciliation of research and education* which Flikkema advocates. In communities of learners, projects are related to the 'big ideas' of the discipline in which the community is situated. Students see the link between their current research or professional work and the tradition in which they operate. This 'situatedness' corresponds with Flikkema's view that academic communities have a *clear sense of heritage and historical perspective*. In communities of learners, students acquire the necessary methodological skills to conduct their research or professional skills. Moreover, they learn to communicate and to collaborate. Gradually, they are introduced into the culture of their discipline.

In communities of learners, individual and collective reflection is practised and used as an important tool of the mind in order to relate concrete experience with abstract theories. Practising individual and collective reflection skills contributes to the development of what Flikkema terms a *sense of doubt*, which *protects students from the tendency to 'form instant opinions'*. In communities of learners, participants can work with resources and equipment which are available in their research environment or in the professional environment of their discipline. Creating communities of learners in universities is clearly in line with Flikkema's vision that universities should be reshaped to make them future-proof institutes that serve the common good and that create *knowledge for life*, as Brink' (2007) put it.

This third knowledge creation metaphor surpasses the first two metaphors of knowledge acquisition and participation; it may well be more resilient and less vulnerable to the push towards profitability. Examples can be found in recent programmes initiated at Vrije Universiteit Amsterdam (Beishuizen, 2008, 2015). One of these, the Cancer Project, offers a perfect illustration of the meaning of the knowledge creation metaphor. A brief description will be given below.

The Cancer Project is part of the second-year Bachelor's programme in Medical Natural Sciences. The project aims to create a better understanding of the origin of cancer: which molecular processes turn a healthy cell into a tumor cell? Students are introduced into radiotherapy, chemotherapy and the characteristics of research in the domain of molecular biology. Throughout the Cancer Project, students collaborate with teachers and researchers to disentangle the characteristics of (potential) oncogenes by manipulating DNA and studying the effects in a yeast model. In turn, their instructors use the outcomes of the Cancer Project to strengthen their own research. Now why is this project a clear and effective example of a community of learners in which students contribute to knowledge creation? Because the data gathered during the Cancer Project are useful for the instructors' research programmes and the students feel that they are seen as serious partners in the process of knowledge creation. In the project, education and research are reconciled because students construct their own understanding by actively being engaged in research. The instructors describe their role as *research based teaching*. The project's core issues are clearly related to the *big ideas* of the discipline of molecular biology. The students are introduced into the culture of conducting research, collaborating and communicating. Students learn to reflect on their work and on the meaning of their research outcomes, particularly during the planning sessions that are held at the start of every project meeting. And last but not least, students work in a fully equipped research lab, supported by PhD students, assistants, analysts and their instructors.

Conclusions

Universities should adopt various educational strategies. The knowledge acquisition metaphor refers to an educational strategy with a focus on the transfer of academic knowledge to students entering the domain of their future career. It seems an appropriate strategy for the early years of Bachelor's programmes, but not exclusively. At later moments, too, students need to acquire substantial amounts of knowledge and learn to master a number of skills to be ready for their professional life. The participation metaphor refers to an educational strategy with a focus on enculturation: adopting the language, culture and habits of the professional environment for which students are preparing. Over time, students gradually develop their professional identity through peripheral participation. This strategy becomes more and more important in later years, not only in Bachelor's programmes but also in Master's programmes.

The knowledge creation metaphor, with its focus on individual and collaborative knowledge development through research, encompasses the knowledge acquisition as well as the participation metaphor. As we have seen in the example of the Cancer Project, students acquire knowledge and skills by actively constructing their understanding of the domain in which they conduct research. Moreover, they participate in the world of research by collaborating in a community of learners and working in a research lab. In fact, the knowledge creation metaphor refers to an educational strategy which can be used as an appropriate tool to realize Brink's (2007) combined purpose of advancing knowledge for life. In this way, both Plato and Aristotle are served – and the university becomes a real School of Athens. I claim that this combined educational strategy creates a stronger position for universities to withstand the profitability push that Flikkema warns us against: stronger not only because this strategy serves both aims of knowledge acquisition and participation, but also stronger because this strategy surpasses these separate aims by enabling and preparing students to create knowledge for their own lives and for the good of society. In this way, students become autonomous, competent and caring academic citizens. Flikkema's university of the future, educating students to create knowledge for life, does in fact resemble the School of Athens.

References

Ausubel, D.P. (1968). *Educational psychology, a cognitive view*. New York: Holt, Rinehart and Winston, Inc.

Beishuizen, J.J. (2004). *De vrolijke wetenschap: Over communities of learners als kweek-*

plaats voor kenniswerkers [*The sparkling science: on communities of learners as incubator for knowledge workers*]. Amsterdam: Vrije Universiteit, Onderwijscentrum VU.
Beishuizen, J.J. (2008). Does a community of learners foster self-regulated learning? *Technology, Pedagogy and Education*, 17(3), 183-193. http://doi.org/10.1080/14759390802383769
Beishuizen, J.J. (2015). *Tien jaar communities of learners in praktijk en onderzoek* [Ten years of communities of learners in practice and research]. Farewell Lecture Vrije Universiteit Amsterdam. Amsterdam: Vrije Universiteit.
Brink, C. (2007). What are Universities for? *Vice Chancellor's Lecture Presented on 27 November*. Newcastle: Newcastle University.
Brown, A.L., & Campione, J.C. (1996). Psychological theory and the design of innovative learning environments. In L. Schauble & R. Glaser (Eds.), *Innovations in learning; new environments for education* (pp. 289-326). Mahwah, New Jersey: Lawrence Erlbaum Associates.
Healey, M. (2005). Linking research and teaching exploring disciplinary spaces and the role of inquiry-based learning. In R. Barnett (Ed.), *Reshaping the university: new relationships between research, scholarship and teaching* (pp. 67-78). Maidenhead, UK: McGraw Hill/Open University Press.
Joseph, P.B. (2011). *Cultures of curriculum*. New York: Routledge.
Lave, J., & Wenger, E. (1991). *Situated learning: Legitimate peripheral participation*. Cambridge: Cambridge University Press.
Nussbaum, M.C. (2012). *Not for profit: Why democracy needs the humanities*. Princeton, NJ: Princeton University Press.
Paavola, S., Lipponen, L. & Hakkarainen, K. (2004). Models of Innovative Knowledge Communities and Three Metaphors of Learning. *Review of Educational Research*, 74(4), 557-576. http://doi.org/10.3102/00346543074004557.
Sfard, A. (1998). On two metaphors for learning and the dangers of choosing just one. *Educational Researcher*, 27(2), 4-13.
Volman, M. (2006). *Jongleren tussen traditie en toekomst* [Juggling between tradition and future]. Inaugural lecture. Amsterdam: Vrije Universiteit.

'The Times They Are a-Changin'[1]

HALLEH GHORASHI

When I arrived in the Netherlands in the late eighties, as a refugee from Iran, I found myself in a very uncertain situation. I didn't know if I would gain permission to stay, and I was far away from all that I knew and loved. Paradoxically, this uncertainty also offered me safety and new opportunities in life. That realization felt like an enormous boost of energy, as if I was given a second chance to live and make new choices in life. I felt like a coiled spring that had been compressed for ages and suddenly, in the Netherlands, had been set free, releasing all its vitality and curiosity. At once I set out to learn Dutch, and a year after my arrival in the country, much to my own amazement, I enrolled in a cultural anthropology programme at Vrije Universiteit Amsterdam. The support of UAF (the Foundation for Refugee Students) was crucial for gaining admission to the university. Yet studying in another language and entrance to the university seemed so unreal that I could not quite believe that this dream had come true until I actually found myself sitting next to fellow students in lecture halls. The students were much younger than I was (I was 27 years old), and the first six months of study were not easy, but I was determined not to waste a single second. Being a student gave me a positive perspective that was the complete opposite of my insecure life as an asylum seeker. My refugee background made me dedicated and determined to turn my new life into a success, but I could not do it alone.

My life as a student also helped me to move beyond the trials

* Halleh Ghorashi is professor of Management of Diversity and Integration in the Department of Sociology at Vrije Universiteit Amsterdam.

1 The title of this essay refers to Bob Dylan's well-known song released in 1963.

of the past, to detach myself temporarily from the most painful memories, and to share the more positive ones. The open ears of my fellow students and teachers gave warmth and purpose to my life. In the first few years of my residency, their curiosity and openness in accepting me as a human being with a story to share was crucial. The small number of students and the level of engagement with each other in and out of the classroom made it possible for me to create meaningful connections. Slowly, but surely, my contacts with others and the opportunity to fully take part in everyday life brought joy back to my life. Small conversations grew into special friendships. I remember vividly that one of our teachers, Professor van Binsbergen, invited us to his home to discuss books together after his course had finished. While not all of the students participated, there was a group of 15 who were there most of the time. We ate the food that we all brought, and while his children played in the room we discussed amazing books and contemporary issues.

As students, we also planned trips to the Ardennes and Center Parcs (which I have to admit were less successful) for midweeks, and those formed great memories. I often invited one or two of my friends who also were refugees to come along on those trips, and the group welcomed these newcomers with warmth. When I think back, I am sure that these *delayed/slow* moments of in-depth interactions and meaningful conversations were crucial in creating a feeling of connectedness and warmth in my university life during those early years in a new country, which would otherwise have been full of insecurity and emptiness. At that time, the university climate facilitated playfulness and reflective engagement within a world of inequality and indifference. Yet now that I have been part of the academic communities in the Netherlands for more than 20 years, I find myself asking this question: *Are our universities facilitating the same kind of playfulness and reflective engagement that was crucial in my (academic) development as a refugee student?*

This is also the central point of my essay: What kind of message are we giving our students about the world and their role in it? How are we going to provide the inspiration for our students to be original and reflective about the challenges that surround them? I find these questions especially relevant in a climate dominated by cynical and negative discourse concerning cultural diversity in Europe, and especially in the Netherlands.

When it comes to the dominant pattern of discussions in Europe, it seems that critical thinking, which requires some doubt of one's own judgement, has been replaced by a reactive attitude – rooted in the fear of loss and change – of protecting one's own culture and identity. Fearing change makes it impossible to remain curious about different points of view. A protective attitude towards one's own identity and culture limits creativity, originality, and does not allow us to be open to change. Fearfulness of change brings us to a situation in which we prefer to maintain the status quo and choose interaction with replicas of ourselves in order to feel safe and secure. Yet, falling victim to this attitude means denying the flow of our time. The world is changing, and a defensive attitude towards this change is counter-productive.

Considering that the core task of academia is to provide social critique and reflection, universities have an undeniable role in the quest for increased inclusiveness and resilience in relation to the normalized structures of exclusion. As early as 1959, C. Wright Mills (reprint 2000) argued that sociology needed to live up to its promise and moral imperative, which in his view was a social analysis that was of direct relevance to urgent public issues and insistent human troubles. He called on sociologists to use what he referred to as 'the Sociological Imagination' to connect everyday personal realities with larger social realities (Mills, 2000: 15). Mills argues that there is a need for narratives and sociological imagination to make a difference in the quality of the human condition. This means engaging with the narratives of uneasiness, indifference, and exclusion by connecting them to historical and societal contexts. I would like to go beyond social sciences and argue that the task of academia is to improve the human condition (Arendt, 1958) and to create conditions for developing meaningful, reflective narratives that bring the past and present together and in so doing move beyond hasty and thoughtless actions in order to enable a sustainable future.

Zygmunt Bauman (2000) argues for academia to come as close as possible to yet hidden human possibilities by going against obvious and self-evident assumptions. The task of critical theory is to question the normalizing power at work through reflection and societal engagement. He argues that in late modern societies it is the invisible forms of power which are dominant rather than visible or coercive forms. In this approach, power is not visible and tangible but works through routinization and normalization of everyday practices. Power, in this sense, is located in the processes which we take for granted and, for that matter, often reproduce even when they are not in our favour. What are the conditions and possibilities for change in these normalized structures? When the sources of exclusion work through normalized and repetitive practices of everyday interactions, the major way to resist is to create delayed spaces for reflection. The act of delay protects us from what Thomas Eriksen (2001) refers to as 'the tyranny of the moment'. 'To go fast means also to forget fast' as Lyotard argues (in Janssens & Steyaert, 2001: 109). The hastiness of our actions strengthens the power of normalization. Stopping to think and creating delayed in-between spaces of reflection in our hurried routines enables us to rethink our own position in these processes and, from time to time, think of actions that can disrupt these taken-for-granted structures.

Generally, the major task of an academic education is to provide students with valuable insights so that they will not take the processes they are part of for granted. Instead, they should keep raising critical and constructive questions, which are the foundation of academic thinking. By doing so, students learn to claim their subjectivity and their power of self-creation. This claim of subjectivity means that they create new meanings, as opposed to being subjected to certain practices and becoming the object of the act. Through exploration and reflection, students are challenged to question their position in the normalizing structures they are part of.

Having said that, it is impossible to enlarge the societal imagination of our students or our society without being reflective about the academic structures we are part of. How can we teach our students to be reflective about their position in society and broaden their sense of connectedness when universities are focusing more and more on individualized excellence and producing quantifying methods (number of publications, number of citations, amount of funding) to measure quality? How can we encourage more societal imagination when universities are complicit in the loss of that very imagination by discussing quality and existence of curricula based on the number of students? Why do we expect society to take our contributions seriously when we are fixated on and encouraged in thinking of our individual career development in a linear, vertical manner? When formal and informal processes and decisions are based on output performances, the process of reflection is neglected as are the rich interactions and delayed spaces, which I argue are crucial for academia. An example of this is the way that the newly celebrated open spaces in our university offices (which are paradoxically closed to outsiders, including our students) diminish the space for the concentration necessary for individual reflection because of the lack of silence while simultaneously limiting the possibility of meaningful conversations because of the production of noise. Caught in-between many such dilemmas, we need to ask ourselves what has happened to our sense of serving society. Are we still meeting the expectations to enlarge the imaginative capacity of society through original research, engaged teaching, and the promise of providing examples of thoughtfulness and humility (the more you know, the more you know what you do not know)? Above all, is there any place left in our universities for vulnerability, solidarity, and humanity when we reproduce structures that praise individual perfection, over-assertiveness, and overachievement as sources of success and almost single sources of promotion?

In conclusion, it is certain that times are changing from my time as a student. The existence of Bachelor's and Master's programmes with small numbers of students (which was crucial for my sense of connectedness) is under serious attack. Delayed spaces for reflection and meaningful conversations within academia (with and among colleagues and students) are diminishing because of the increasing work pressure based on the often impossible expectations of performing excellently at all levels. In addition, it is useful to consider the *Matthew effect*: once one receives a prize or a prestigious research grant, the rest will follow. How fair is this system and how dignified are our actions and interactions with valuable colleagues who have invested enormously in the process and do not win the prize? How good are we in considering hidden talents or excellence that do not fit the dominant linear criteria of success, and are for that reason invisible? How realistic is the new emphasis on societal impact when it is expected that academics do that in addition to all the other things they have to do? All of this leads to the increase of work pressure in academia: working days and nights and even weekends and feeling the (in)formal pressure of being available at all times to perform when needed. In order to survive, many of us are just contributing to these normalized structures by working harder and harder with the hope that at some point our talent will gain notice and we will fall into the loop of the *Matthew effect*. Becoming a visible success does not necessarily make us wiser and certainly does not leave us enough time and space to delay, to reflect and to inspire our students and our societies through engagement. What is left of this process are a few celebrated academics that have been awarded prizes without having time to spend the money and a large group of academics who are frustrated and who have lost their motivation and their self-confidence to explore their talents and to be inspiring forces in academia. As academics, we need to question these normalizing structures that remove the breeding ground for playfulness, imagination, and reflective engagement in academia. Without this breeding ground, we cannot serve our students and our societies as sources of inspiration to enlarge their imagination.

Our university, Vrije Universiteit Amsterdam, is one of the few which does not shy away from its commitment to society but has formulated a strong ambition of serving society as one if its core values. Is this not the perfect time to redefine the meaning of *Vrij* to bring it closer to the challenges of our time? In its new meaning, *Vrij* would refer to being free from the normalized structures that put system values (such as productivity, output, acquisition, number of students) above professional values (such as originality, innovative thinking, imagination, humility, collegiality, and integrity). Freedom in this sense for our university means not to be one of the many other institutions choosing the ivory tower of excellence, but to become the example of the university we claim to be: a university that invests in individuals, in meaningful conversation, in thoughtful expressions, in dignity, humility, and delayed spaces of reflection and engagement. A university that invests in processes instead of solely praising positive outcomes. A university that is able to improve the human condition, by looking back to be able to look further. Only such a reflective university can contribute to durable solutions in an era when the foundations of a sustainable future (both natural and social) are seriously challenged.

References

Arendt, H. (1958). *The human condition*: University of Chicago Press.

Bauman, Z. (2000). *Liquid modernity*: Polity Press.

Eriksen, Thomas Hylland (2001). *Tyranny of the Moment: Fast and Slow Time in the Information Age*: Pluto Press.

Janssens, Maddy and Chris Steyaert (2001). *Meerstemmigheid: Organiseren met verschil*: Universitaire Pers Leuven.

Mills, C. W. (2000). [1959] *The sociological imagination*: Oxford University Press.

An Educated Guess

JAMES KENNEDY

It is a great privilege to address you today[1]. I do so, though, with great trepidation and humility, especially realizing how many of you have dedicated your lives and your talents to the education of our students. As for myself, especially in this medieval setting, I feel a bit like Dante the Pilgrim in the Divine Comedy as expressed in its famous first line: 'In the middle of the journey of our life I came to myself within a dark wood where the straight way was lost'. My motivation to enter into this profession was to teach – in the first, second and third place. Gradually, though, I found myself doing less and less of it, focusing more and more on other opportunities and responsibilities, including research (which I experienced as enabling my teaching but also in practice in partial tension with it), administration as a well as a host of other activities outside of the university. Today I am myself, and perhaps like many of you, a seeker, looking to make 'educated guesses' about how we can, to the best of our insights and abilities, educate, that is, to 'lead forth' and to 'draw out', students for their work in the world.

That task, as you know, is not easy. As fundamental as teaching and learning are to the life and the purpose of the university, we often suspect that they are not as central to the mission of universities as they ought to be. In fact, universities are often places where only quite 'limited learning' takes place, to cite Arum's and Roksa's research of American universities in their book *Academically Adrift*

* James Kennedy is Dean of Utrecht University College and professor of Modern Dutch History. From 2003 to 2007 James Kennedy worked as professor of Modern History at Vrije Universiteit Amsterdam.

1 Keynote speech at the Onderwijsparade 2016 (University Utrecht).

(Arum and Roksa 2011). They paint a highly critical picture of a university system in which virtually all actors had reasons for not pushing learning processes further. Students in the United States seek to 'acquire the greatest exchange value (a degree) for the smallest investment in time and money.' Academic staff preferably spend their time pursuing their own research and other professional interests, and administrators are focused on finances and university reputation. Government agencies chiefly find the university interesting as a place where new scientific knowledge is generated. Good teaching could still make a difference, Arum and Roksa found, but it did not happen to the extent to relieve their generally somber picture. Though their findings are controversial – some critics point to evidence that critical thinking skills really do improve in college – and not necessarily applicable in all respects to the European situation, it seems to me clear enough that there are structural patterns in place that gravitate against the kind of teaching from which both 'excellent' and 'average' students would benefit most.

Despite such assessments, there is of course much good that has happened with education in the university. More than in the recent past teaching has been recognized as important. And for over twenty years this university has worked hard to improve the quality of its education and to emphasize the importance of this education. Commitment to the quality of university teaching as a point of policy was still something of a countercultural stance in the 1990s; it no longer is. We know that from the professoriate at this university on the basis of surveys that their estimation of teaching and its importance has risen significantly in recent years. Moreover, the Center of Excellent in University Teaching (CEUT) has empowered many teachers to learn from each other, enhancing their teaching in the process.

And yet there are limits to these positive developments. The problem is not only, or perhaps even chiefly, the proverbial precedence of research over teaching in the effective hierarchies of the university, which leave a good many teachers systematically undervalued despite some modest efforts to reverse this. It lies in at least two other developments that I see at this university. The first is that educational programs – more than ever – have been subjected to tighter controls, controls incidentally mostly meant to improve the quality of education and to guarantee some kind of public accountability for educational 'product.' That education be well managed, coherent, transparent and indeed accountable have been increasingly important concerns. This raises the real possibility that the dimensions of teaching – not easily subject to the rules of accountability because too 'subjective' to be so evaluated – have possibly stagnated for lack of encouragement, because these aspects do not fit into what is formally required of us as a university. In short, educational programs seem to be much more about creating coherent curricula and measureable outcomes than about creating room for developing the hearts and minds of our students (to paraphrase Ken Bain in *What the Best College Teachers Do*, Bain 2004).

And the second related issue is that this university still does not interact sufficiently with the wider world. I am aware that some important initiatives have been undertaken – I am somewhat familiar – and charmed- with the *wetenschapsknooppunt* that connects UU students with local schools. But the centrality of connecting *teaching* to a wider public task has not been systematized. Research has been making a turn to pay more attention to 'societal impact', but the educational wing of the university arguably has lagged behind, in part, I think, because of deep ambivalence in the academy about whether such an aim is properly part of the university's teaching mission. It seems more sensible and perhaps safer to define 'learning outcomes' according to more ostensibly measureable academic criteria.

I sense that there is a growing discomfort with this tightly circumscribed educational system, as the protests at my former employer the Universiteit van Amsterdam illustrate. After years of increased emphasis on accountability and '*rendementen*' there is some willingness to ponder, certainly also at this university, whether all of this has gone too far, and that we should be 'let go' a bit, loosen the reins as it were, when it comes to education. This emerging reflection should not, however, be restricted to questions whether particular rules should be relaxed or particular 'learning outcomes' broadened or abandoned. Much of the problem we face is deeper. Our assumptions about teaching are premised on the idea that 'learning' is not only to be effectively measured but is also to be guaranteed – at least as long as certain well-defined processes are set into motion and certain outcomes can be demonstrated. But it is very much the question whether this is the case. Inspired by Gert Biesta (whose recent *The Beautiful Risk of Education* (Biesta 2013) is often the source of insight in my presentation today), I tend to think that learning, by students or anyone else, is not a given. Teaching, moreover, as William James once argued and Biesta affirms after him, is an art – not a science – with often wonderful and many times unforeseen results (James 1925). And because it is an art education at the same time carries attendant 'risks' that therefore carries in it the very real possibilities of failure, of *not* being able to lead forth, of *not* being able to draw out, or from the student's view, of not being led forth or not being drawn out. Teaching and learning are dialogical processes subject to radically different 'outcomes' (if I dare use the word), even between good students and good teachers. Biesta reminds us in a deeply human endeavor such as education that 'input' will never totally correlate to 'output.' Something like 'evidence-based' teaching still needs to take account of millions of different contexts in which only the art of the teacher can hope to find paths forward. I think that this is an important and humbling insight for everyone – for the Ministry of Education, for university administrators and not least for teachers themselves, who always have to recognize that the ability to teach is not something that is always given to them.

So there are compelling reasons why 'letting go' and leaving room for the 'educated guess' might actually prove beneficial to the way we further develop our own thoughts and practices about education. It seems to me that there are two ways to bring us to a better place, though they might be in partial – but perhaps also creative – tension with each other.

The first requires considerable attention to, and giving room to student engagement. Contemporary higher education has ceased to engage the heart, Tim Clydesdale of the College of New Jersey asserts, while active learning really requires 'existential engagement' (as Mark William Roche of the University of Notre Dame argues in *Why Choose the Liberal Arts?*) that links inner concern and passions of the students with the needs of the world (Clydesdale 2015; Roche 2010). That's not always easy to accommodate. Sometimes students don't either know how their studies relate to their commitments, or they do not seem to care how they might be. A majority of American students seeks a degree in higher education to relate to a 'higher' purpose of their lives, but that percentage is down from a generation ago. And of course the research university itself feels ambivalent about matters of 'existential engagement' or of linking university teaching to 'social responsibility' which Andrew

Delbanco of Columbia University sees as one of the main aims of a university education (Delbanco 2013). If 'existential engagement' is restricted to possessing a love for specialized research, then we all feel quite comfortable; this is, after all, exactly *Wissenschaft als Beruf* as Max Weber articulated it. But necessarily tying that deep love of research, as crucial as it is, with a wider mission, of concern for the world, is much more difficult. Few at the university, to be sure, want to produce merely amoral technocrats, though there remain a significant number of academicians who do think, like the American literary theorist Stanley Fish, that students should 'save the world on their own time'. I think it is more common that many academicians are just uncertain about to what extent they should undertake it and perhaps just how they might. Added to this hesitancy perhaps is a pious hope that a wider social or moral engagement of students will take care of itself. It may – but it also might not.

To address this problem there have been important initiatives to put a new emphasis on student motivation. This has taken many forms. One such initiative are 'learner-led' programs that seek to put students more in 'control' of their own learning processes by giving much more room for them to design, at least to some extent, their own courses and program of study. In this way they can identify their own interests and their own set of concerns. This is just one approach; there are others, of course. The humanities program at this university, including University College Utrecht, work with portfolios which ask students to reflect on what they learned and how this might relate to the wider world. I think this kind of reflection is *very* important, although I am still looking for the proper form. I observe sometimes that compulsory written reflections that by their nature cannot be graded can be deadening for both teacher and student unless both happen to be 'in the mood' or are eager to engage in such exercises. Maybe we can find ways to do this that are less bureaucratic and less prone to a going-through-the-motions mentality that is the exact antithesis of real reflection.

However we may think of portfolios, reflection does matter. It might be worth mentioning that years ago when I was working in the United States I was part of a team that received a large grant to develop 'purpose exploration programming' for the students at our college, in which students could sign up for a range of programs to discover their passions and tie it in with their curriculum and their further aims. In this way they could, as one source puts it, 'thoughtfully calibrate their life trajectories during their college years' and develop a 'grounded idealism' (Clydesdale 2015). Such projects continue to be financed in many U.S. colleges and universities; one of them has developed select programs that foster (in two separate initiatives) 'urban engagement' and 'leadership development.' Although diverse in ambition and scale, all really are about stimulating thoughtful discernment among students about the future lives, and linking that with exploratory actions on their parts. Though not everyone who goes through such programs finds it rewarding, many do, and those who have a stronger sense of meaning and purpose in their lives than those who did not.

In various ways, universities have come to recognize the importance of the internal motivations of students. Our university certainly has, seeking as it now does to heighten 'student engagement' in all faculties, which sometimes has included the creation of co-curricular 'communities' in which students and teachers can explore wider themes and issues in ways that enhance a student sense of embeddedness, and of engagement in their own studies, apparently with some success. And this letting go can be quite far-reaching at University College Utrecht; the course molecular cell biology, under direction of Johannes Boonstra and Fred Wiegant, chiefly consists in allowing students to write their own high-caliber research proposal, compelling them to do their own original research to make this possible. The expertise of the instructors is critical in several ways for this approach to work, but it is a great example of how teachers can *trust* students to find their own way to achieving a very high standard of work.

A particular focus of my own is to find ways of enhancing civic engagement among my own students, not just only in doing good for the community (as laudatory as that is) but as a way to link their own academic interests with wider needs. In the United States, 'service-learning' projects have been established in hundreds of universities, which links course work with partnerships in civil society and with government to achieve some public good, all the while giving room for students to become further motivated by their contacts outside the 'ivory tower'. It is something I would like to see more of here, though it already has its analogs here.

Seen this way, helping students deepen their own internal motivations for studies is a crucial way for enabling us to let go. If we can trust more in their own motivations and their own drives then perhaps we can permit ourselves to do less box-checking – all the while keeping in mind that deep forms of participation will also carry risks. One of the tasks here today, then, is to think about how we can create the conditions for students to explore their own motivations and the kinds of letting go that such endeavors might entail. Some of my concerns would be that such plans would not really substantially change the way that the university works, that is, that student engagement is fine as long as it fits within the existing imperatives and structures of the university. And universities, being deeply conservative institutions, are going to want to do that.

And there's another concern, too, and this brings me to my second point: that to the extent that we open things up to students we actually might help shut things down. Biesta, himself an inveterate critic of the neo-liberal developments in education, rightly insists that students are not, should not be seen as, 'consumers whose needs need to be met' (Biesta 2013, 57) – an important critique that we should always use to measure our own motivations for improving education at this university. He is also critical, as I also happened to be in a 2004 address (which after many years I looked up to prepare for today), of the teacher (merely) as 'coach' or as 'facilitator' who may or may not be important to the learning process, and where 'learner-led education' is the dominant model (Iversen, Pedersen, Krogh & Jensen 2015). Biesta's critical stance toward a learning-based philosophy as opposed to a teaching-based one is partially based on the insight that too much emphasis on 'learning' can promote pedagogical solipsism in which a student only can learn from what s/he gets out of herself or himself, rather than seeing learning as an adventure in which one opens oneself up to deeper ambitions, wider perspectives and interest in the insights of the teacher. He has been a firm proponent of 'giving teaching back to education,' and that real education requires an openness from students to new insights from 'outside.' For him, 'being taught' – and not 'learning from' – is essential to education (Biesta 2013, 44-52).

That teachers still matter for university learning was revealed by a poll held last year by Gallup in the United States. Only a small minority of students (one in seven) recalls university as a seminal experience, but those who did, felt more satisfied not only about their work but about their lives. At the very top of the list of factors that made all the difference was the experience of a teacher that made a student excited about learning; it also made a difference to many students if there was a teacher who cared about them, or was willing to serve as their mentor. A third factor – a long-term/semester-long (research) project was also important to many students. In short, if you actually had some or all of these experiences as a student – care, mentoring, guidance in a big project – you were much more likely to think more highly about where you have wound up. Whatever the reasons for this pattern, it is clear that teachers matter. Interestingly, it didn't seem to matter very much what kind of *school* students went to, whether big-name or less celebrated institutions, liberal arts colleges or large state universities. It was the relationship with the teacher that was decisive. To put it more strongly, it is about the ability of the teacher to engage her or his students, so that they may experience the university as a seminal, transformative experience ('Life in College Matters for Life After College' 2015).

It is, I would stress, not the teacher as omniscient and authoritative being that is being smuggled back into the equation, but a teacher who knows that education is a risk; a risk that requires perhaps more than anything creating the optimal kind of pedagogical ethos. That ethos, in my mind, must be one that affirms the kind of virtues that are essential for academic work. These virtues include, amongst others, a self-critical stance to what is learned and taught. This self-critical stance disciplines ourselves to grapple with material that goes against what we would rather see, or at times simply seems tedious. So, without creating an ethos in which attitudes and dispositions such as self-discipline and hard work but also curiosity and openness to new perspectives are instilled, scholarly and scientific endeavor would come to a standstill. Additionally, though, this ethos includes a strong invitation to link the intrinsic motivations of students to the academic work that lies ahead. That may mean a 'culture of care' as Meindert Flikkema at the Vrije Universiteit has argued, in which all teaching must be deeply mindful of the human relationships that form the essence of the teaching-learning process.

For myself, I have been most inspired by the idea of Leon Kass of the University of Chicago who saw 'thoughtfulness' as the aim of a university education. For him, such education was 'the cultivation in each of us of the disposition actively to seek the truth and to make the truth our own. More simply, liberal education is education in and for thoughtfulness. It awakens, encourages, and renders habitual thoughtful reflection about weighty human concerns…' – in whatever discipline, it might be added (Kass cited in Hickerson 2010).

It's worth thinking about 'thoughtfulness' as key to several different ideals of the university. It is a key virtue for friendship, and the recently-retired provost of Valparaiso University, Mark Schwehn, has opined that 'academies at their best can and should become communities where the pleasures of friendship and the rigors of work are united' (Schwehn 1993, 61). But thoughtfulness as a key quality of academic life does other things as well. It offers room for reflection so necessary a companion to the restless drives of our students. It is moreover a reminder of Michael Oakeshott's claim that university education is 'liberal' in that the university is a place of learning, and of being taught, where we do not have to have concerns about the instrumental value of what is learned, or what has been taught.

Attempting to live out the right kinds of dispositions, helping to stimulate an ethos that optimally serves students, and make all of this integral to everyday teaching, is, though, not a path to certain pedagogical triumph, but an error-prone search. To be sure, awareness of, and experience in, particular teaching techniques can be useful means to our aims. But in the end, teaching remains, and must remain, an educated guess. That is why it is all the more important to heed Biesta's call to give more room to the teacher as professional, to permit him or her, as he puts it, to become 'educationally wise.' He argues for a kind of 'virtuosity' in teaching that places emphasis on the wider formation of the teacher, on the freedom to develop teaching practices of one's own, to develop through experience one's own virtuosity and the key role of mentorship in inspiring the teacher further along. Seen from this perspective, the art of teaching as acquired wisdom rather than as the master of technique, or the guarantor of results, is what is critical.

'Letting go' means, then, two things. It means that the organization must let go, in that it relents more in letting the professional teacher develop her or his own virtuosity. But it also means that the teacher herself must let go, and understands that teaching is at heart a 'risky' venture, that it is good and necessary that it is – and that letting go of control may make more things to happen than before.

The emphasis that I have placed here on two facets – the critical importance of intrinsic student motivation and the freedom of the teacher to teach – can be in tension with each other. But they are also related. They are related to each other in the more superficial sense that they both challenge an educational system that, all professions and good intentions aside, is not yet sufficiently invested in either. But they are also bound up in the question of what education actually is for. For both student and for teachers, it must mean more than jumping through the hoops of a disciplinary study. It requires a deeper, more ambitious engagement with the material and a deeper sense of personal calling that can only be fostered in a particular kind of community, where teaching really is considered the most important thing we do at the university and where the motivations of students are, to the fullest extent possible, both encouraged and challenged. What the result will be will necessarily be an educated guess – but as stakeholders in education and particularly as teachers – we surely wouldn't want to have it any other way.

References

Arum, Richard, & Roksa, Josipa (2011). *Academically Adrift. Limited Learning on College Campuses.* Chicago: University of Chicago Press.

Bain, Ken (2004). *What the Best College Teachers Do.* Cambridge: Harvard University Press.

Biesta, Gerd (2013). *The Beautiful Risk of Education.* Boulder and London, Paradigm.

Clydesdale, Tim (2015). *The Purposeful Graduate. Why Colleges Must Talk to Students About Vocation.* Chicago: University of Chicago Press.

Delbanco, Andrew (2013). *College: What is Was, Is and Should Be.* Princeton: Princeton University Press.

Hickerson, Michael (2010). 'Thoughtfulness as the Aim of Liberal Education?' http://blog.emergingscholars.org/2010/08/thoughtful-as-the-aim-of-liberal-education/).

Iversen, Ann-Merete, Pedersen, Anne-Stavnskaer, Krogh, Lone & Jensen, Anne-Aarup (2015). 'Learning, Leading, and Letting Go of Control'. *Sage Open.* October-December, 1-11. http://sgo.sagepub.com/content/5/4/2158244015608423.

James, William (1925). *Talks to Teachers on Psychology. And to Students on Some of Life's Ideals.* New York: Henry Holt and Company. http://www.uky.edu/~eushe2/Pajares/tt1.html.

'Life in College Matters for Life After College' (2015). http://www.gallup.com/poll/168848/life college-matters-life-college.aspx.

Roche, Mark William (2010). *Why Choose the Liberal Arts?* Notre Dame: Notre Dame Press.

Schwehn, Mark (1993). *Exiles from Eden. Religion and the Academic Vocation in America.* Oxford: Oxford University Press.

Back to the Drawing Board: Do Universities Need a Redesign?

ELCO VAN BURG

'An aircraft is only a piece of aluminum. As a pilot am happy to fly it, but for us as an organization it is an indispensable means to an end.' This is how our experienced chief pilot – with an impressive number of 10,000-plus flying hours – reflects on the new airplane that has just arrived. The date is September 2015, and finally, after having been forced to wait for seven months, our small social venture *Lentera Papua* in the rural highlands of Papua (Indonesia) is up and running again. The plane is essential: not only for our pilot training programme for the benefit of local Papuans, but also for the cash-flow that is needed to keep all the social services up and running. With his words, our pilot highlights the fact that we need to keep our goals in mind. In running an aircraft operation, remembering the aim of training locals and providing other social services is often a challenge, as the business usually requires full attention. Yet during the past few months, we have been able to reflect on our mission and aims. Since the business came to a standstill in January, staff have stayed on – even without wages – and wanted to talk about why we are doing this and what our key values are. Although the period was stressful and uncertain, the end result was that we found confirmation that *Lentera Papua* truly is the project that we would like to work for: we aim to train local Papuans to be dedicated and mission-driven professionals in the field we work in, and we are willing to pay the price it takes for doing this.

When I was back at the university for a couple of months, to

* Elco van Burg is consultant at *Lentera Papua*, Papua (Indonesia) www.lentera-vanburg.nl and part-time employed as associate professor in entrepreneurship at the Faculty of Economics and Business Administration at Vrije Universiteit Amsterdam.

teach, supervise and conduct research, I started to reflect on this period with *Lentera Papua*, triggered by Meindert Flikkema's thought-provoking essay 'Sense of Serving' (Flikkema, 2016). Although I realize that the idea may be rather foolish, I wish that there was something we could do at the university that resembles our Indonesian project: stop certain processes for a while and think about why we are here – and what we want to achieve. Typically, when I am at work at the university, it feels more like being overloaded by competing and sometimes even conflicting demands rather than having sufficient time for reflection. Students ask all kinds of questions about exams, books and assignments. PhD students send drafts that need my input. I find myself running to meetings, preparing lectures while travelling on the train, and the end of the day often leaves me dissatisfied that I have not been able to work on the five research projects that are sitting on my desk. At the same time, however, it is this combination of tasks and responsibilities that makes my work meaningful. Still, I am glad to have the luxury to be able to look at universities from a distance every now and then, literally and mentally, when I am in Indonesia. In this essay, I shall build on these reflections with the aim to put on the drawing board some of the design principles for universities, and in particular the discourse within universities.

Debates about universities

A few years ago, when I was still full-time involved in academia, I wrote a dissertation on university spin-offs, targeting the topic of – forgive me for using a Dutchism – 'research valorization' (Van Burg, 2010) and subsequently becoming involved in the discussion on 'selling the university' (Van Burg, 2014). As a PhD student, but especially in my role as research coordinator in our department, I learned about an issue that was much more pressing for large groups of university staff: the demands for research output had become high, leading to situations in which research has to compete with education in terms of available time (this key tension is also emphasized by Flikkema, 2016). Most people, however, actually seem to stress the reverse: teaching demands are eating up our research time, while at the end of the day, when it comes to promotion decisions, we are evaluated on our research output.

In an academic reflex, I started to look for answers to these problems in the academic literature. Yet, delving into popular as well as academic literature commonly consulted at universities, the first things I came across were additional debates about universities, research and education. First, I noticed the popular press and journal editors complaining about research *quality* and pointing out shortcuts designed to obtain long publication lists: plagiarism, salami publishing, data fabrication (cf. Martin, 2016) and other questionable research practices (cf. O'Boyle, Banks & Gonzalez-Mulé, 2014). Relatedly, researchers and funding institutions are increasingly often questioning research *relevance* (cf. Van de Ven, 2007), and researchers lament that real breakthrough work and innovative research are hindered by academic conventions and promotion procedures (cf. McMullen & Shepherd, 2006). Second, especially after the recent economic crisis, people have started to question the quality and relevance of university *education*, including education at the top business schools which had educated the managers that unintentionally laid the basis for the economic crisis (cf. Romme, 2016). More in general, the entire system of ranking academic institutions – including managerial attention for these rankings – is currently being critiqued (Adler & Harzing, 2009), partly because these rankings suggest that all universities are animals of the same breed, although in actual practice a top school like Stanford University and a largely unknown institute in Indonesia are as different as chalk and cheese.

These debates are as lively in Europe as they are in the US, and they may even be louder in the US. To illustrate, just take a look at all the books that have been published with titles like *University Inc.*, *College for Sale*, *The University in Ruins*, *Selling the Ivory Tower*, *In Defense of American Higher Education*, and *Wannabe U.* One review of the latter work nicely summarizes many of the issues that academics often complain about (Stevens, 2010, p. 1042): 'There is the gradual but relentless growth in numbers and titles of administrators. There is the obsession with measured admissions inputs, academic outputs, and institutional rankings. There is the overlay of organizational and environmental change on intergenerational faculty succession, such that senior faculty, with their purportedly obsolete conceptions of university life, are doomed to codger status. There is the large and pervasive importance of courting big donors. There is the chronic contraction of state support for the university and the constant hunt for new revenue streams. And there is of course the 'wannabe' phenomenon itself – the capacious prison of middling status in which countless ambitious schools and their personnel are sentenced to endless, unflattering upward comparisons.'

Designing the discourse

Reflecting on these debates and perceived issues for universities, predominantly in the western part of the world, I would like to take the liberty to stop for a little while and go back to the drawing board. I think that this is helpful, at least as a thought experiment, to get a clear vision of what it is that we want to achieve at the university and how we want to do it. The multiple debates – as outlined above – involve multiple tensions and possible design choices. That said, as a pragmatist, I am interested in the question what we can do about it in terms of crafting solutions together with all the stakeholders involved rather than in terms of describing an ideal vision or model for a university that needs to be implemented. In general, I believe that devising blueprints or *ideal types* for universities (for instance the ideal type Humboldtian university) is not really helpful in the professional community that a university forms – apart from the thought-provoking function that such ideals can have. Instead, drawing on design thinking in organization design as a reflective practice (Schön, 1984), I propose developing a set of guiding design principles that help to design the discourse about what the university could become.

In such a design science approach, the key parts to be specified are design elements and design principles. Design elements describe what can be altered in the design without changing the class of the object being designed. Design principles describe possible interventions that lead to a certain outcome or set of outcomes, and sometimes add an explanation of how these interventions lead to the outcomes. In management, steering away from 'fixed designs' and archetypal thinking, design thinking is amongst others applied to develop heuristics for effectual decision-making in entrepreneurship (Sarasvathy, 2004) and to describe practical design principles for corporate venturing practices (Van Burg, De Jager, Reymen & Cloodt, 2012).

To develop fully-fledged design elements and principles, a systematic review of the literature as well as codifying managerial practice is recommended. This essay is not the appropriate place for me to develop a complete set of design principles and present a systematic literature review. Moreover, key to designing discourses is that this requires a bottom-up approach, without predefined outcomes in mind. This means that design elements and principles need to be specific enough to guide the discourse in order to make it relevant and to the point, but at the same time they need to be sufficiently generic to avoid predefining the direction of the discourse within the academic community, or *universitas*. Nevertheless, building on a set of review and overview papers available in the literature as well as general design science insights, I *can* present a preliminary set of design elements and discuss aspects that need to be included in design principles in order to help design the discourse about the university. As such, these design elements delineate the topics that need to become subject of the discourse.

1. The first design element is that of a shared vision for a university. To facilitate a fruitful discourse among professionals, there needs to be some form of shared vision or imagination (Romme, 2016). For universities, this means that the university community and management need to agree, at least to some extent, on what the future of the university should look like, given the context that the university is in (see Barnett, 2011). For many larger universities, in particular public universities, this will very likely take the form of a 'pragmatic vision' rather than a utopian ideal type (Badley, 2014).
2. A second key design element is the type of governance that fits this vision (cf. Trakman, 2008). The more specific the vision, the more explicitly university governance can be steered towards this vision. The design principle(s) for this element need to specify how to deal with the so-called *New Public Management* reforms that have been implemented in many public universities and that are focused on increasing efficiency in public organizations (cf. Christensen, 2011) and on measurable output such as publications (Flikkema, 2016). Moreover, HRM practices need to be defined (Musselin, 2013) that contribute to the vision that we propose (to illustrate: some research performance measurement systems do indeed increase research output, but they also reduce diversity and societal relevance, see Hicks, 2012).
3. With its dependence on vision and its effects on governance, the element of teaching quality, methods and approaches is a third

item to be considered (cf. Díaz-Méndez & Gummesson, 2012). This includes how the university – or departments within the university, if they have the autonomy to decide on these topics – deals with digital learning materials and environments such as massive open online courses (MOOCS).
4. Regarding the element of research quality, topics and approaches, design principles need to give guidance on how to organize research, how to fund research and how to evaluate research (cf. Hicks, 2012).
5. A final element concerns the way in which interactions with society and the economy are formed. Here, the design principles need to give guidance on how to deal not only with technology transfer, university-industry relationships (cf. Perkmann *et al.*, 2013) and university spin-offs (Van Burg, Romme, Gilsing & Reymen, 2008), but also with the question how the regional role of the university can be fulfilled.

Imagine the university

The role of the university in modern society has significantly changed over the years. Approximately fifty years ago, a university education was something for a highly selected group of people, but this situation has changed significantly. In the Netherlands today, for instance, half of the Dutch thirty-year-olds holds at least a Bachelor's degree; in the US, this number stands at 32%[1]. The enormously increased role of university education alongside revolutions in research practice and changes in society and the economy make that 'old' models of universities may no longer hold – or at least fail to offer the utopia that we want to build with our current universities. Therefore, we need to imagine new futures: not just one future, but multiple futures, and we subsequently need to make careful, joint choices about what we want to do – while simultaneously staying flexible and open to change, new insights and new opportunities. In this respect, it is of key importance that universities form professional communities in which the community as whole – including students, academic and non-academic staff

1 'Nederland wordt steeds slimmer', DUB. Retrieved from www.dub.uu.nl/plussen-en-minnen/2014/09/15/nederland-wordt-steeds-slimmer, November 21, 2015; 'Educational Attainment in the United States: 2014'. U.S. Census Bureau. Retrieved from www.census.gov/hhes/socdemo/education/data/cps/2014/tables, November 21, 2015.

as well as management – can engage in joint sense-making of the imagined university. Here, design thinking can help us to shape potential development trajectories and to make deliberate choices about each of their design elements. In essence, this concerns a joint endeavour and a joint responsibility, so that we may shape the university of the future. After all, in the academic community everyone is responsible – although perhaps not always to the same extent – for the end result.

References

Adler, N.J., & Harzing, A.-W. (2009). When knowledge wins: Transcending the sense and nonsense of academic rankings. *Academy of Management Learning & Education, 8*(1), 72-95.

Badley, G. (2014). The pragmatic university: A feasible utopia? *Studies in Higher Education*, 0(0), 1-11.

Barnett, R. (2011). The idea of the university in the twenty-first century: Where's the imagination. *Journal of Higher Education, 1*(2), 88-94.

Christensen, T. (2011). University governance reforms: Potential problems of more autonomy? *Higher Education, 62*(4), 503-517.

Díaz-Méndez, M., & Gummesson, E. (2012). Value co-creation and university teaching quality. *Journal of Service Management, 23*(4), 571-592.

Flikkema, M. (2016). *Sense of Serving.* In: M.J. Flikkema (Ed.), *Sense of Serving – Reconsidering the Role of Universities in Society Now*. Amsterdam: VU University Press.

Hicks, D. (2012). Performance-based university research funding systems. *Research Policy, 41*(2), 251.

Martin, B.R. (2016). Editors' JIF-boosting stratagems: Which are appropriate and which not? *Research Policy, 45*(1), 1-7.

McMullen, J.S., & Shepherd, D.A. (2006). Encouraging consensus-challenging research in universities. *Journal of Management Studies, 43*(8), 1643-1669.

Musselin, C. (2013). Redefinition of the relationships between academics and their university. *Higher Education, 65*(1), 25-37.

O'Boyle, E.H., Banks, G.C., & Gonzalez-Mulé, E. (2014). The chrysalis effect: How ugly initial results metamorphosize into beautiful articles. *Journal of Management*, forthcoming.

Perkmann, M., Tartari, V., McKelvey, M., Autio, E., Broström, A., D'Este, P. & Sobrero, M. (2013). Academic engagement and commercialisation: A review of the literature on university–industry relations. *Research Policy, 42*(2), 423-442.

Romme, A.G.L. (2016). *The Quest for Professionalism: The Case of Management and Entrepreneurship*. Oxford, UK: Oxford University Press.

Sarasvathy, S.D. (2004). Making it happen: Beyond theories of the firm to theories of firm design. *Entrepreneurship Theory and Practice, 28*(6), 519-531.

Schön, D.A. (1984). *The reflective practitioner: How professionals think in action*. Aldershot: Ashgate.

Stevens, M.L. (2010). [Review of *Review of Wannabe U: Inside the Corporate University*, by G. Tuchman]. *American Journal of Sociology, 116*(3), 1042-1043.

Trakman, L. (2008). Modelling university governance. *Higher Education Quarterly*, 62(1-2), 63-83.
Van Burg, E. (2010). *Creating spin-off: Designing entrepreneurship conducive universities.* Eindhoven: Eindhoven University Press.
Van Burg, E. (2014). Commercializing science by means of university spin-offs: An ethical review. In: A. Fayolle & D.T. Redford (Eds.), *Handbook on the Entrepreneurial University* (pp. 346 369).Cheltenham, UK: Edward Elgar.
Van Burg, E., De Jager, S., Reymen, I.M.M.J., & Cloodt, M. (2012). Design principles for corporate venture transition processes in established technology firms. *R&D Management*, 42(5), 455-472.
Van Burg, E., Romme, A.G.L., Gilsing, V.A., & Reymen, I.M.M.J. (2008). Creating university spin offs: A science-based design perspective. *Journal of Product Innovation Management*, 25(2), 114-128.
Van de Ven, A.H. (2007). *Engaged scholarship: A guide for organizational and social research.* New York: Oxford University Press.

Tragedy at the Modern University: An Advocacy for Bildung and Participatory Pedagogy

TINEKE ABMA

Grandmother Trijntje

My grandmother Trijntje, after whom I was named, was born in a tiny little village in the north of Friesland. She got married, ran a bakery and a family with four children. Hers was a poor family, and only on Sundays – after church – could they afford a special meal. Although they had not much to share themselves, each Sunday afternoon my mother would go to see four poor neighbours. For all of them, Trijntje had cooked some extra pudding with pears: for the old, crippled lady, for the lady with tuberculosis, and for ''the two old unmarried brothers'. Every week, Trijntje also managed to save a few coins for the VU's medical centre. She taught me what it means to serve other people and society, and she continues to be a source of inspiration to this very day.

Today, our society has become hyper-individualized, rationalized and economically driven (Bauman & Donskis, 2015). Although prosperity has increased, living and sharing – serving – like Trijntje did in her days is not so obvious anymore. This societal development is reflected in the way in which our universities are organized. They have become focused on efficiency and output, now more than ever, rather than on character building, identity formation (determining what really matters to a person) and moral considerations. System norms have become dominant and are putting professional and moral values under pressure. This has fuelled a broad-based sense of discontent about our universities, and calls

* Tineke A. Abma is professor of Participation & Diversity and vice-chair at the Department of Medical Humanities, Vrije Universiteit medical centre Amsterdam.

for a fundamental reorientation on the question what purposes the university serves.

In this essay I will first describe the crisis at the university in terms of the split between research and education, competition as the way to excellence, mass education and a lack of democracy. These problems, I will then argue, are grounded in a much more fundamental crisis, namely the alienation of moral responsibilities. In today's postmodern world, this is an urgent matter. If we want to prepare students and young researchers for life, teach them to be good citizens and academics who feel morally responsible, we need to integrate cultural and moral learning processes in education and research. Serving universities make room for *Bildung* and engagement and nurture these through dialogue.

Competition as a way to excellence?

In this spring's student protests against the budget cuts in the Humanities Department at the University of Amsterdam, resistance against a focus on efficiency and productivity (In Dutch: 'rendementsdenken') became manifest (Thomas, 2015). The protests gained vast support among teachers and researchers, and highlighted a broader set of problems currently inhibiting and negatively influencing the quality of our academic practices, here and elsewhere in the world.

First of all, a clear split has emerged between research and education, tasks that used to be intimately integrated. Especially researchers have profited from this separation: in a monomaniacal fashion, they can focus on research without being concerned about education and ways to share their knowledge and competences with students. The split makes it more complicated to translate research into education, and researchers are becoming increasingly less skilled at disseminating knowledge and supervising young people. As a result of the high status given to scientific research, we now see a hierarchy within the university between science and education.

Secondly, within the domain of research, a competition model has been introduced in which money has to be acquired via a system of proposals and review procedures. This stimulates entrepreneurship, but it has also led to a situation in which researchers spend almost one third of their time writing proposals of which only a small part gets funded. The focus on output (number of publications, PhDs and acquisitions, for instance) puts major pressure on people, creates a climate in which academic integrity is undermined (consider data manipulation, for example, or salami slicing techniques to produce separate publications that should have been submitted as a single manuscript), and values quantity above quality. The quality of researchers is counted and measured in terms of scientific impact by means of quantitative instruments. These were introduced into the world of research by managers to monitor the performance of researchers and institutions. Their focus is on citation scores in journals with certain impact factors. One of the most frequently used scores is the so-called Hirsch-index, which combines citation impact and productivity. A high H-index is considered to be an effective indicator of being a 'good' researcher.

The above system focuses on individual researchers, and does not reflect or acknowledge the complexity and collective character of scientific research. Commonly, the design, completion and implementation of research findings requires collaboration among researchers with different talents, in different stages of their careers. The bibliometric system cannot accommodate this type of collaborative work. On the contrary, it fuels competition, because every researcher will do his or her best to get the highest possible H-index. In these conditions, investments in matters that are important from the perspective of the collective, such as building relationships and networks as well as taking time for intellectual exchange and deliberation, are not stimulated. Furthermore, it leads to a situation in which cultural-critical approaches and the societal relevance of research beyond its purely economic valorization are undervalued. Competition has furthermore resulted in flexible contracts for almost sixty percent of the workforce, creating uncertainty and tension among staff. Another effect is that PhD students are solely trained to conduct a piece of research and to write articles; they hardly develop other academic skills such as teaching, critical thinking and implementing knowledge.

Mass education and democratic deficit

A third issue to be considered concerns numbers: more people attend the university than ever before. In itself, this is a good trend, because the number of highly educated people goes up. Yet, this has also led to mass education, and to more distant and impersonal interactions; direct contact and interaction between teacher and student is scarce. Today's student generation has other expectations of teachers and does not automatically accept their expertise. Students have grown up within a more horizontal culture of negotiation instead of a more hierarchical culture, and they expect teachers to be open and willing to discuss matters with them, including their grades. In addition to this 'horizontalization' of relationships, students have grown up in an information age in which knowledge and expertise are easily available on the Internet. Where an individual's expertise and authority were easily accepted in the past, nowadays many expert opinions are available side-by-side. As different, and sometimes conflicting, expert opinions are ventilated, the notions of truth and expertise become contested. The teacher is therefore no longer automatically accepted as an authority, and has to gain – and be granted with – authority on the basis of trust. Moreover, in a hyper-culture of fragmentation and instant consumption ('I want it all, and I want it now', to cite Freddy Mercury) students expect teachers to be performers entertaining them with videos, stories and other catchy materials.

Not all teachers are willing and equipped to handle these expectations. These teachers are unwilling because they know that good teaching is more than satisfying students. Learning often includes perturbations and confusing disruptions, which are not always pleasurable when they occur, but of great importance in the longer run (Niessen, 2003). And not everything that students have to learn is of direct use. Teachers therefore rightly emphasize that they have their own professional values and norms, and that these have to be cherished and nurtured. Unfortunately, no substantial budgets are available for training teachers to maintain these professional values and norms and to develop themselves, personally and professionally. Teachers have been neglected for many years, they do not have many career perspectives, and they are set increasingly higher output targets. Although there are still some people who combine research and teaching, many focus on one of these domains. Due to the hierarchy in research and teaching, less attention has been paid to those who are doing the teaching. As a result, innovations in higher education and attempts to improve its quality are lagging behind.

Finally, many students and teachers complain about the limited degree of their participation in policy design and decision-making processes within university. This democratic deficit can be understood as follows: formally, students can become members of a student council. This is supported by legislation. Although student councils have some rights, in practice many feel that they have become mere bystanders with little influence on strategic decisions. Students are approached as 'consumers.' Their *choice* is important, and the array of educational programmes is increasing in order to give them more options, but their *voice* hardly matters to academic policymakers and executives. The recent protests have shown that students do in fact want to have 'a say', but that executive management is too distant and hardly open to new voices and perspectives. From a management perspective, the protests and resistance are considered troublesome and not constructive, instead of a welcome impulse to innovate (Van der Ven, 2008).

History tells us that the democratization of universities resulted from student revolts, but since the '70s, student rights have been limited step by step for the sake of efficient decision-making. In the '80s, the new Dutch act on university governance, the 'Wet Universitair Bestuur' (WUB act), was adjusted to give the executive boards of universities more power. In the '90s, this law was further transformed into a modernized version, the 'Modernisering Universitaire Bestuursstructuur' (WUB act). Under the WUB act, practically all strategic decisions had to be approved by the University Council ('Universiteitsraad') consisting of students, teachers and other employees. Today, student councils are still legally entitled to approve some issues, such as the rules concerning education and examinations, but in many instances students only have the right to give advice. Thus, student councils can still exert influence, but this is not based on legal entitlements; rather than anything else, their influence is the effect of the mobilization of other students and employees. The recent student protests have shown that social activism based on ideals is still alive, and that students as 'competent rebels' gather and use other, more deliberative and direct forms of democracy to express their opinions (Thomas, 2015). This kind of 'disobedient politics' is characterized by a non-hierarchical organization based on temporary coalitions and leaders, very much like the Occupy and other square movements (Van Gunsteren, 2015).

On a more fundamental level, the instrumentalization of teachers (in creating a maximum output or 'ECTS' per department and doing this as efficiently as possible), of students (the more the better), and research (meant to generate or obtain cash funding, with publications in high-ranking journals seen as the 'currency' of academics) affects our self-understanding as researchers and teachers. In fact, all this is grounded in financial argumentations and interests. More students, for example, means more money, but it does not necessarily lead to better education in terms of quality. Likewise, more PhDs generate more money, but this says nothing about the quality of young researchers.

The instruments and techniques currently used to assess the output of research and education appear to be neutral and innocent, but in fact they communicate and endorse technocratic and market values – i.e. related to efficiency, safety, control and managing – and expert knowledge, i.e. the expertise of professionals, management and policy makers, but *not* the values held by students and other societal members. As such, this mode of thinking in terms of measurement and performance has a greatly significant impact on university practices and the people involved. Practitioners working with these techniques are more and more led to think about their practice, their own role, and their identity in instrumental ways. Education, for example, is increasingly often understood in terms of a 'market' with 'consumers' – seen as highly autonomous beings in making decisions about their lives – who need to be satisfied with certain 'products' or 'services.' Today's clock-time rationality with its focus on short-term outcomes does not acknowledge the fact that students have their own personal rhythms of learning (Sabelis, 2015). System values and norms also dominate research, leading to a situation in which productivity – generating as much output as possible – is more important than devoting time to 'slow questions' that cannot be solved instantly and that require critical thinking and careful deliberation.

Such an instrumental approach has little to do with what it means to be a good teacher or a good researcher or what it means to provide good education to students. When teachers are invited to count and measure their work, to enhance factors such as their productivity and student satisfaction, they are implicitly geared to become more proficient at controlling and managing situations, and not at learning to teach students to think critically. This approach does not acknowledge the complex, relational and unpredictable character of the profession (Van Manen & Shuying, 2002), and undermines motivation as it ignores the initial vocational call and motives why teachers and other professionals opted for their work in the first place. The spirit, wisdom, energy and time needed to meaningfully engage students dries up when one's work is controlled by systems and measures that cannot reflect its nature and values that are truly important, such as altruism, caritas, solidarity, empathy and care, but also critical thinking, reflection and deliberation. Thus, the world becomes disenchanted; substantial values become mere abstractions, and functional rationality becomes dominant. In this way, practitioners become dissatisfied and drift away from their actual daily work environments, where knowledge and insight grow contextually and where knowledge and insight are *contextually* meaningful (Snoeren, 2015). Practitioners no longer need to think about the goodness of their work, nor what good education or research actually means in a particular situation for a particular student. Things are considered 'good' when they meet pre-ordained targets and evaluation criteria (student scores, for example, or publication numbers). In other words, instrumental techniques discourage practitioners from thinking about their practice in moral or normative terms and how they can be held morally accountable for the decisions they make. This leads to alienation: alienation from moral responsibilities. Here, not only the Stapel case springs to mind, but also the manipulation of student scores and cases of management fraud we have witnessed in higher education. These recent examples of amoral behaviour are by no means incidental. They are part of a deeper moral crisis in the university and in culture in general. As Bauman and Donskis (2013) aptly put it: 'A consumerist attitude may lubricate the wheels of the economy, but it sprinkles sand into the bearings of morality' (p. 150).

System values and norms are putting increasing pressure on professional norms held by teachers and researchers. This not only undermines their passion and motivation, but it also has consequences for the system itself. The entire academic system is based on professional values and personal engagement on the part of teachers and researchers. Whether it is giving attention to an apparently disinterested student or listening to a young researcher who suffers under the pressure to publish, delving up their talents and encouraging them, all of these values are seen as some sort of 'extra' from the perspective of the system, while in fact these form the very *heart* of teaching and research. They cannot be endlessly exploited by the system, but they need care and nurturance to flourish and survive.

Bildung and participatory pedagogy

If we want to prepare students and young researchers for life, and if we want to teach them to be good citizens and academics who feel morally responsible, we need to integrate cultural and moral learning processes into the curriculum and research tenure tracks. This is particularly important in a postmodern society in which morality has become highly individualized and taken over by the market. Moral guidance is no longer embedded in overarching and mutually shared frameworks, ideologies and grand narratives (Bauman, 1993; Kunneman, 1998). This postmodern crisis calls for a re-personalization of morality and urges citizens to take on moral responsibility instead of relying on abstract rules and principles. Moral responsibilities and values are, however, not just given, but need training and development. We have to teach students to understand the moral ambiguity of certain situations and to deal with the consequences of their choices while being aware of the interdependence with others. Ethics starts from the assumption that people (should) have a deeply entrenched and deeply felt consciousness that everything is connected with everything else. Naturalistically 'felt' responsibility emerges from the realization that one inhabits and shares a relational universe. How can educators support themselves and others to come to this deep realization?

Serving others starts with trying to understand each other and, to that purpose, engaging in dialogue. The philosopher Hans-Georg Gadamer argued that dialogical processes of understanding lead to *Bildung*: (moral) development or cultivation (1975). Gadamer stands in a long tradition of philosophers who argued that *Bildung* should be central to academic education. Von Humboldt's nineteenth century idea(l), revived in the 1960s by Jürgen Habermas, was that universities take upon themselves the task to promote civilization and make world citizens out of their students with open minds, not afraid to judge and think for themselves. Gadamer considers venturing into the position of others through dialogue to be crucial to this process. He even defines *Bildung* as 'trained receptivity to otherness', and argues that not thinking about one's own standpoints, concerns and interests for once, but rather making an empathetic and honest attempt to understand the other, brings about a transformation of the self through which one cultivates oneself. Dialogue therefore needs to be at the center of our academic praxis, to bring about moral learning processes and a *joint* and *moral* understanding of what good teaching and research entails.

While *Bildung* stimulates critical reflection and moral learning, students also need to learn to act for change. Particularly in a city as Amsterdam, home to many minorities, it is important to understand value pluralism and power asymmetries and appreciate how these lead to structural disadvantages. What we need is a 'participatory pedagogy' of teaching and learning that results in action-oriented self-understanding (Freire, 1972, 1990). This particular pedagogy can also stimulate and include research practices that are both societally relevant and critical. Participatory research is an activity in which participants engage and collaborate in a process of researching their own lives and communities, resulting in action-oriented learning and understanding (in terms of self and collective). This is a context-bound, complex and dynamic process integrating various social groups and various sorts of knowledge (Abma *et.al.*, 2010). It is through the embeddedness in these often messy practices that the researcher may truly encounter others (Cook, 1998) and engage in fruitful dialogues with them.

Epilogue

The modern university is increasingly driven by economic powers and strategic behaviour, and research and teaching run the risk of becoming all too sterile practices. The tragedy of this focus on efficiency and productivity is that it leads to alienation of moral responsibilities, demoralization and disenchantment with our academic life, which is then ironically countered by the same modern 'recipe' of even more instrumentalization (Frissen, 2005). An illustrative example is the bibliometric system that is applied to determine the social value of research but which redirects the focus on measurable academic output.

It is needless to say that in the above context the bottom-up and intrinsically motivated initiative of Sense of Serving is of utmost importance and urgently needed, because Sense of Serving, in essence, is a moral-ethical appeal to sensitize students, teachers and researchers to the needs of others and topical societal issues. This closely resembles Trijntje's efforts when she responded to the people around her who had found themselves in vulnerable situations and when she supported higher education and healthcare for everyone. This was partly a matter of duty, but also of deep concern with and moral responsibility for the development of young people and the care for those in need of help.

Sense of Serving requires more than the mere generation and transfer of knowledge. It requires *Bildung*, the moral development and cultivation like that offered by the teacher in the cinema film *Dead Poets Society*. The teacher provokes his young students not to just accept what the school system and their parents obliges them to learn, but to follow their curiosity, to step out of the utilitarian worldview, and to collectively read and discuss poetry. This means putting dialogue and participatory pedagogy at the centre of our praxis. In the actual context, true encounters with others – those who ask indiscrete questions, who are vulnerable or helpless – can take place. This confrontation may put us off centre, but it will stimulate reflection on rational schemas and representations, and ultimately create a moral appeal on our part to serve the other. This precedes and exceeds rationality, and it can offer us the chance of learning to live meaningful lives. It is therefore an answer to the broadly felt 'disenchantment' with our world.

A serving university enables *Bildung* and engagement through dialogue. It acknowledges that the academic system can flourish thanks to the rich interactions, moral deliberations, personal engagement as well as the professional and moral values of teachers and researchers. In our postmodern world, these professional and moral values are urgently needed and cannot be endlessly exploited by the system. A serving university nurtures the normative professionality of teachers and researchers. Their values and their engaging interactions with students and societal partners are not some sort of luxury or extra, but form the *heart* of professional quality in the academy: they form the basis for professional action.

References

Abma, T.A., V. Baur, B. Molewijk & G.A.M. Widdershoven (2010)., Inter-ethics, Towards an interactive and interdependent bioethics, *Bioethics* 24(5): 242-55.

Bauman. Z. (1993). *Postmodern Ethics*. Oxford: Basic Blackwell.

Bauman, Z. and L. Donskis (2014). *Moral Blindness: The Loss of Sensitivity in Liquid Modernity*. London: John Wiley & Sons.

Cook, T. (1998). The importance of mess in action research, *Educational Action Research*, 6(1): 93-109.

Freire,P. (1972/1990). *Pedagogy of the oppressed*. Continuum: New York.

Frissen, P.H.A. (2005). *De staat van verschil. Een kritiek van de gelijkheid*. Van Gennep, Amsterdam.

Gadamer, G.H. (1975). *Wahrheit und Methode. Grundzüge einer philosophischen Hermeneutik*, in: *Gesammelte Werke*. vol.1., Mohr: Tübingen.

Kunneman, H. (1998). *Postmoderne moraliteit*, Boom, Amsterdam.

Niessen, T. (2007). *Emerging Epistemologies, Making sense of teaching practice*, Datawyse Maastricht, Maastricht.

Sabelis, I (2015). Reflections on academia from a perspective of time(s). This text is downloaded on June 19, 2015 from *Open! Platform for Art, Culture & the Public Domain onlineopen.org/article.php?id=475*

Snoeren, M. (2015) *Working = Learning. A complexity approach to workplace learning within residential care for older people*, Ridderprint BV, Ridderkerk.

Thomas, C. (2015). Competente rebellen. Hoe de universiteit in opstand kwam tegen het marktdenken, Amsterdam University Press.

Van der Ven, N. (2008). *Schaamte en verandering, Denken over organisatieverandering in het licht van de filosofie van Emmanuel Levinas*, Klement, Kampen.

Van Manen, M. & L. Shuying (2002). *The pathic principle of pedagogical language, Teacher and teacher education*, 18: 215-224.

Acknowledgment

I thank my colleagues Suzanne Metselaar, Susan Woelders and Petra Verdonk for their valuable feedback on earlier versions of this essay.

‘Eppur si muove: The Earth Revolves Around the Sun’ (Galileo)

STEVEN TEN HAVE

Probably misattributed to the Irish poet William Butler Yeats but nevertheless true, ‘education is not the filling of a pail, but the lighting of a fire’. Meaningful education is more organic than mechanical, dynamizing rather than stabilizing. It is not about doing a deed, but getting a movement going. In the field of education, fire has plenty to offer: not only as a metaphor, but the analogy also offers insights. When humans learned to control fire, they progressed from their ecologically secondary position to being ecologically dominant. Comparably, education elevates mankind to a higher order. Fire brings heat, light and protection. It enabled hunter-gatherers and the first farmers not only to control pests, but also to create open spaces and new possibilities. Education is capable of doing something similar; there was good reason why Mandela saw education as the most powerful weapon in the struggle to create a better world. In the movement that has to ‘live’ up to this, ‘sensing’ should be the ultimate product of ‘knowing’.

Since the educational road through lessons is a long one while the path through examples is short and effective, we'll consider the case of an exceptionally intelligent man called Will and his instructor Sean in the cinema film *Good Will Hunting*. There is one fragment that perfectly illustrates the difference between ‘sensing’ and ‘knowing’. Will ‘knows’ everything, but ‘comprehends’ little. Sean says: ‘If I were to ask you about love, you'd hit me with a sonnet. But you've never looked at a woman and been truly vulnerable ... feeling like God put an angel on earth just for you. Who could rescue

* Steven ten Have is professor of Strategy and Change at Vrije Universiteit Amsterdam and professor by special appointment at Nyenrode Business Universiteit in Breukelen.

you from the depths of hell?' Similarly, meaningful academic education is not effectively characterized by the analytical distinction between *Bildung* (general education) and *Ausbildung* (vocational education), burdened by the link with the distinction between the elite and the masses, with conservatism or Nietzsche (1883) and his concept of the Übermensch. Meaningful academic education implies an integrative understanding of what mankind is, means, can do and produces. More succinctly, it provides students with perspectives on the meaning of mankind.

'Sensing' embraces 'knowing', not because of having it or knowing it, but originating from conceptual essence. This embracement becomes manifest in 'stories'. The psychologist Jerome Bruner (1987) teaches us that a message is twenty times more likely to be remembered when it is conveyed through a good story than when it is based on facts and figures. This holds particularly true for personal and interpersonal stories which resonate with, inspire and light the audience. We also want to mention what Ibarra (2015) calls conversion stories and journey stories. These are archetypical stories that we hear when we grow up: biblical stories, fairy tales and legends. They represent conversion experiences: moments when it all snaps into place and after which nothing is the same again. The one event changes everything. These 'moments of truth' or 'critical' life-changing incidents do in fact exist in many lives in every culture and religion. But from a learning perspective, the 'journey stories' may be even more important and more relevant if we focus on the way in which learning really 'works'. Ibarra (2015) mentions the metaphor to be found in the story of Ulysses on his long and wandering journey to Ithaca, a quest filled with many challenges, uncertainties, turns, 'crises', deep experiences and temptations to stray. The 'profit' – all or part of it – lies in the process, an enriching and formative journey. As Robert Frost[1] puts it: 'We'll get lost along the way, lost enough to find ourselves'.

That said, our education system is not characterized by journey stories. In his *Pedagogy of the Oppressed*, published in 1970, Paolo Frere describes what continues to remain the dominant model of teaching today. Students are viewed as empty 'bank accounts' to be filled with the knowledge of teachers. They are not seen or treated as participants, let alone 'co-creators' who have a 'voice' in

1 In Ibarra, 2015, p. 159.

what to learn and how to learn it. This model is not about learning and inspiring, but about assessing and controlling. It is not about thinking and adapting; it is about absorbing and stabilizing. In the prevailing education system, learning processes, access to information, assumptions, paradigms and methodologies are precooked in a closed loop. This possibly serves or perpetuates existing social structures or power hierarchies. It provides a kind of guarantee, certainty and comfort, and is sometimes motivated by pointing at the 'civil effect' of diplomas. However, it also positions and designs learning and education as a production or throughput process instead of a search and development process, or a process of self-actualization.

Let us go back to academic education, and learning, as lighting a fire. Meaningful academic education recognizes the importance of combining knowledge and skills, personality, talent, fascination and socialization, and self-realization. If fascination and the passionate will to develop and learn are fire, then knowledge is fuel. Heat represents serving, and self-realization is oxygen. Academic education is often considered from the point of view of utility or consequentialism. The alternative that fits our message is supplied by March en Weil (2005): the fulfilment of our identity, 'our sense of self'. This also clearly indicates that in education we allow ourselves to be ruled by consequences. We study, teach, control and 'stimulate' out of sheer necessity, because 'we' or 'they' think that this will lead to 'good' results. The focus seems to lie on careers, money, quality, accreditation and the realization of policies and party platforms. The extrinsic dominates the intrinsic. Dan Pinker (2010) teaches us that intrinsic factors are essential in processes such as the learning process. People long for autonomy, progress, mastery and purpose. People who are motivated to learn and teach want to lead their own lives, they strive to become better at something that truly matters, and they yearn to contribute to something that is larger than their own selves, for the sake of the communities in which they live.

Locked in our systems and structures, we think that the sun revolves around the earth. Because of our fixation on consequences, we forget what is truly desirable and effective: to 'bring on' the good, to fulfil our vocation. Bringing things on, irrespective of consequences, leads to an identity with consequences and to consequences with identity. This is both 'instrumental' and 'valuable'.

It will help and qualify for living life: living it from a humanistic perspective. This is important and necessary in developing academic education in the right direction. If we want to improve and develop academic education, we need to define 'good' and 'better'. As a consequence, we have to talk about values and valuation or appreciation, about intentions, goals and purposes. Decisive interaction is interaction that deals with the educational system's purpose and the identity of stakeholders who really care about their students. Theoretically, a system's goal may range from drilling people to letting them flourish or recreate themselves. Identity can be defined both from a technical, transactional and an adaptive, transformational perspective. Education itself can be characterized by knowing (alone) or by sensing. What we need to do is answer the question what the purpose of education is or has to be (for us). This means that values such as defining and qualifying elements are absolutely critical.

Meaning can be found where purpose and shared identities in our society truly meet in a harmonious and humanistic way. Then, meaning is brought about when we are willing to see what really moves people and when the freedom of movement does not exist in freedom alone, but also in safety. In addition to offering space and autonomy, we therefore need to provide learners with context and containment. Both context and containment become manifest in providing structures, giving direction, paying attention, being there, guaranteeing a 'fair process', giving love and seeing each other. Whatever the form, safety is non-negotiable. Taking context and containment seriously leads to another, different language that does what it is supposed to: make contact, reach out and 'speak' with each other. In addition, it will help if concepts such as *Bildung* are substituted by attachment. Attachment revolves around safety and offers the basic principle for valuable academic education. Attachment stimulates a sense of being approachable, the ability to enter into relationships, to learn, to build self-confidence and develop a 'sense of self'.

With attachment, we enter not only the sphere of the student, but also the sphere of the teacher. Without underestimating the increasing importance of e-learning and other innovations and facilities, the teacher may very well be termed the most crucial external or 'other' part in the educational process of students. It is the teacher, stupid! Or more precisely: it is the relationship between teacher and student. One may see analogies with the importance of relationships in therapeutic settings, the quality of these relationships, and the therapeutic alliance or helping alliance. This means that the quality of the teacher and the ability to interact, help and care make a particularly important difference. The eminent psychologist Edgar Schein (2009) elaborates on helping and sensing. He emphasizes that helping starts with determining what the real problem is that someone is facing. The *humble inquiry* is his way to operationalize sensing; as a helper, it is in your interest to say to your student 'Let's talk a little more. What do you think the problem is? What have you tried already? What would you like to achieve?' This approach has vital consequences for the teacher and his or her attitude. According to Schein, the self-image of the helper as a competent expert is a trap. In the position of a helping teacher, it is the student's problem and related needs, questions and ambitions that decide whether or not the teacher is an expert. The danger for teachers lies in applying knowledge that is irrelevant to the real issue. Knowing, knowledge and technical solutions will not do the job. Sensing, building trust and meeting adaptive challenges require something else. If someone wants to be truly helpful, they have to establish a relationship and a helping alliance, like therapists do.

Caring and helping are not the exclusive domain of our primary schools, as is often thought. Caring and helping are needed in higher education, too. We should wish for a climate that encourages well-prepared students to sit on the edge of their seats, respectfully challenging professors and vice versa. We should foster a system that stimulates students to reflect and experience theory, to write and to inspire others. This should be done as a matter of course, without asking for it as a deliverable. The interaction between tacit and explicit knowledge will bring students new knowledge and generate compelling questions about the system's boundary conditions. The progress that students achieve will drive the pursuit of personal growth. However, what we currently see is that many students sit back, do only the minimum amount of work that is needed to pass exams, and blame professors for being either too demanding or insufficiently entertaining. What has gone wrong and where did it go wrong? Why do professors fail to succeed in lighting a fire within these students? Or, why does the fire die so soon?

The academic workplace has become individualized and dehumanized, and education has become a balancing item. In addition, the unity of research and education is at stake. The ultimate consequence is that providing the foundations for a caring society, a claim still heard throughout today's academic world, has become an illusion. It is espoused theory; it is not theory in use. Current attempts to balance education and research implicitly confirm the competition between education and research, or they are thwarted because academics try to decompose education quality. However, the alienation process we are currently witnessing can be turned around from within, through creating a sense of serving and developing an eco-system that fulfils the potential of knowing (Flikkema, 2016). Servant leadership, reinvented double-loop learning, gratitude from students' high-quality mentoring relationships will serve as components of such a vital system that will foster living together. To illustrate, the mentoring relationships are promoted by attentiveness towards each other and a caring attitude, alongside learning-focussed values (Snoeren, Raaijmakers, Niessen & Abma, 2016). Snoeren *et al.* (2016) state that these relationships are characterised, among other things. 'by person centeredness, care, trust and mutual influence, thereby offering a situation in which mutual learning and growth can occur' (p. 3). The vital system is not primarily an economic system, but a social system that will provide us with meaning for society at large. This is what defines our journey, and we should wish for a guide like Antoine de Saint-Exupéry who teaches us that 'If you want to build a ship, don't drum up the men to gather wood, divide the work, and give orders. Instead, teach them to yearn for the vast and endless sea.'

The crucial leadership challenge related to academic education these days is not a technical but an adaptive one. Harvard professor Heifetz qualifies treating adaptive challenges like technical problems as the single biggest failure of leadership. The solution of technical problems is often characterized by an improvement in current practices. Adaptive challenges, however, are about the disparity between values and circumstances and ask for a deeper questioning of fundamental assumptions and values. This means that the mere application of current 'technical' know-how and routine behaviour is destined to fall short. Instead, tolerance for uncertainty and the presence of divergent voices are essential. One has to be able to *sense* one's 'surroundings', to grasp and understand the 'why' instead of focusing on the 'how'. However, in education one tends to use technical solutions such as formal training, 'quality systems' and policy making to respond to adaptive challenges related to changing societal, cultural and human conditions.

Human beings are no 'empty bank accounts' to be filled, and our educational systems should not be mental and bureaucratic prisons in which people are locked up. Academic environments are meant to be vivid communities of learning and development, of passion and imagination, connecting people around rightful purposes and just causes.

References

Bruner, J. (1987), *Actual Minds, Possible Worlds*, The Jerusalem-Harvard Lectures World

Flikkema, M.J. (2016). Sense of Serving. In: Flikkema (Ed.) *Sense of Serving: Reconsidering the Role of Universities in Society Now*. Amsterdam: VU University Press.

Frere, P. (1970). *Pedagogy of the Oppressed*, New York, Bloomsbury.

Heifetz. R.A. (1994). *Leadership Without Easy Answers*, Boston, Massachusetts, Harvard University Press.

Hall, A.M., Ferreira, P.H., Maher, C.G., Latimer, J., Ferreira, M.L. (2010). The influence of the therapist-patient relationship on treatment outcome in physical rehabilitation: a systematic review. *Physical Therapy*, 90(8), 1099-1110.

Ibarra, H. (2015). *Act like a leader, think like a leader*, Boston, Massachusetts, Harvard Business Press.

March, J.G. Thierry Weil, T. (2005). *On Leadership*, London, Profile Business.

Nietzsche, F. (1883). *Also sprach Zarathustra. Ein Buch für Alle und Keinen*, Chemnitz, Schmeitzner.

Orlinsky, D.E., Ronnestad, M.H. & Willutski, U. (2004). Fifty years of psychotherapy process-outcome research: Continuity and change. In: M.J. Lambert (Ed.) *Handbook of psychotherapy and behaviour change* (5th Ed.). New York: John Wiley & Sons.

Pinker, D.H. (2010). *Drive: The Surprising Truth about what motivates us*, London, Canongate Books.

Schein, E. (2009). *Helping: How to offer, give, and receive help*. Berrett-Koehler Publishers.

Snoeren, M. M., Raaijmakers, R., Niessen, T. J., & Abma, T. A. (2016). Mentoring with (in) care: A co-constructed auto-ethnography of mutual learning. *Journal of Organizational Behavior, 37*(1), p.3-22.

Meaningful Education to Nurture Servant-Leaders

SYLVIA VAN DE BUNT & FONS TROMPENAARS

Servant-Leadership: an introduction

In recent years, the term 'servant-leadership' has become more and more popular, but what does it actually describe? Servant-Leadership (SL) is a leadership principle embedded in a way of life that has been recognized and expounded upon in all parts of the world since ancient times, and across all cultures. In its simplest form, SL is driven by the motivation to enable others to work more effectively and be successful, and can be recognized through active listening, authenticity, stewardship, humility, empowerment and community building, amongst others. SL thus means leading a group or organization in the service of people: clients, customers, employees, students, or members of your network. The more you serve, the more you lead your fellows. Integrating both head and heart, SL expands into the principle of serving a community and acting as a steward of the environment and all that inhabit it (Trompenaars & Voerman, 2009). In brief, the essence of servant-leadership is enabling others to perform better.

The main argument here is built around the fact that servant-leadership works across cultures because it reconciles the meta-

* Dr. Sylvia van de Bunt is Program Director of the VU Servant-Leadership Program (SLRP) and is co-director of SERVUS. She coordinates VU MSc courses in Cross-cultural Management and Careers & Organizations and is visiting professor at two British universities. More information on SLRP and SERVUS can be found at www.feweb.vu.nl/SERVUS.

* Fons Trompenaars is professor of Cross-cultural Management and is co-director of SERVUS. He is the Founder of Trompenaars Hampden-Turner Intercultural Management Consulting, which has recently been acquired by KPMG. Fons Trompenaars is regularly rated on the Thinkers 50 as one of the world's most important management thinkers.

dilemma of serving and leading. One of the main competencies a servant-leader needs to master is to recognize the difference between a problem and a dilemma. A dilemma is a situation where one must choose between two options that seem to be in conflict with each other. This belief leads to the insight that all dilemmas in all cultures are the same, and that the only difference is the starting point and path that we each take to resolve them. For this reason, servant-leaders are not tempted into making a choice between two opposite values, and they also avoid choosing an unsatisfactory compromise. This approach contrasts with one-dimensional linear problem-solving approaches. Once a dilemma has been recognized, the servant-leader helps to reconcile it. He or she connects the team with the individual, uses exceptions to improve the rule and pushes products to be in tune with clients and to listen to them. Hampden-Turner (2016, p.21) argued that the social responsibility of servant-leaders lies in being dedicated to power *through* people not *over* them. Servant-leaders are in the business of empowering followers to succeed them and passing the torch to these followers. According to Hampden-Turner, servant-leaders '*serve meanings which they regard as more important than themselves and that will leave a legacy for mankind.*' In this chapter, we will highlight the SERVUS case and consider how meaningful education can contribute to the development of future servant-leaders.

The origin of the Servant-Leadership Research Programme

In 2010, Vrije Universiteit Amsterdam saw the launch of the Servant-Leadership Centre for Research and Education (SERVUS) with the ambition of examining SL from a historical and cross-cultural perspective. SERVUS' goal was – and still is – to develop professional as well as personal leadership capacity, to help leaders resolve their key dilemmas. The SERVUS mission is in line with Vrije Universiteit Amsterdam's core values (VU, 2015). These three core values serve as a guide to the work and actions of VU's employees and students: being responsible, being open and being personally engaged. Recently, Vrije Universiteit Amsterdam formulated and incorporated its societal responsibility in a Community Service Programme. In this programme, SERVUS acts as one of its collaborative partners (VU, 2016). SERVUS functions as a meeting point for academics and practitioners from various disciplines with the aim to contribute to innovative research and education, and to create value for the community at large. Practitioners from industry are encouraged to engage in dialogue with the academic world, and vice versa. To date, SERVUS has explored SL across various themes, including SL across different periods of history, cultures, disciplines, education, generations and the virtual workplace.

We distinguish three key aims in which SERVUS can play a pivotal role in contributing to the realization of the above, building on Vrije Universiteit Amsterdam's fundamental *raison d'être* as a research and learning institution:

1. To fill a recognized gap in scientific research on the subject of servant-leadership and push the boundaries of knowledge in the field of business studies. Various Fortune 500 firms (2016) and so-called Best Companies to Work For have actively emphasized this approach throughout their business practice;
2. To develop serving (re)searchers' and educationalists' talents;
3. To create an international platform which connects and builds on global (re)search and theoretical developments that are taking place with respect to Servant-Leadership.

To illustrate SERVUS' mission, we will highlight one of the latest SERVUS initiatives, in which we strive to make 'business meet science': the Servant-Leadership Research Programme (SLRP). How do you educate future servant-leaders in a meaningful way? The underlying vision of SLRP is inspired by the desire to develop leaders who are able to serve and engage a meaningful workplace. In the design of such a meaningful educational programme, we were inspired, amongst others, by Edgar Schein's latest work entitled Humble Inquiry (2013), in which he shows how humility in the organizational culture can help create more meaningful conditions for work and learning. Schein's vision is confirmed by Google's Senior Vice President and HR Executive Lazlo Bock. He considers humility a leadership attitude of stepping back and creating space for others to contribute. This attitude is what Bock calls *intellectual humility*; without humility, you are unable to learn (Prime & Salib, 2014). Schein (2013) promotes an organizational culture of humble inquiry; the fine art of drawing someone out and asking questions to which you do not know the answer already. Humble inquiry is the gentle art of asking instead of telling, of building a relationship based on curiosity and interest in the other person.

Dame and Gedwin (2013) describe how humility inspires loyalty, helps to build and sustain group cohesion, fosters productive teamwork, and decreases staff turnover. They illustrate how one motivational CEO once had the courage to give a speech and speak of his own failures, weaknesses and blind spots, and how these had spurred his learning and success. The fact that he spoke about himself in this way deeply impressed the group. Similar motivational speeches are embedded in the postgraduate SLRP organized in conjunction with SERVUS partner De Nieuwe Poort (DNP) based in Amsterdam. At DNP, motivational speakers with the courage to address their failures become personalized role models for meaningfulness and humility. Subsequently, these speakers represent the kind of humble leadership that local companies try to develop for their own organizations while bridging theory and practice. This is the kind of novel dialogue between motivational speakers, faculty members and executives that SLRP envisages in order to nurture meaningful education.

After a period of intensive dialogue with industry, local communities and academia, the SLRP post-graduate course was launched in November 2014. Dutch and Belgian companies nominated their executives who played a key role in the organizational governance of their company to participate in SLRP. Over the past two years, both parties, i.e. participating companies and university representatives, have co-created this course. This intensive dialogue gave birth to the shared vision that meaningfulness and, subsequently, character building should become important dimensions of this Servant-Leadership course. The course was crafted in such a way that field research through humble inquiry became one of its fundaments. Investigating how Servant-Leadership can bring meaning to companies goes far beyond what is common in economic market-driven models. Leading a company is not just about making money; the power model with a primary focus on efficiency is *old school*. A community or organization will only be better off when it creates wealth (Hampden-Turner & Trompenaars, 2015, XIII). SLRP intends to bridge the academic-practitioner gap (Rynes, Giluk & Brown, 2007) by creating a unique combination of the latest insights on Servant-Leadership, community projects and in-company business experience. This reciprocal combination of theory and practice provides applied research-based outcomes for implementation in the participating companies and subsequently for impacting their company's organizational governance.

The SLRP journey

The SLRP journey spans a period of eight months, running from the end of October until its completion at the end of June, with a total of six days of classes and intervision spent at Vrije Universiteit Amsterdam supplemented with fieldwork to be carried out in a community and an in-company project. The SLRP key benefit is primarily to reframe leadership from within the quality of human relationships through class dialogue and fieldwork, all of which takes place in a cross-company team of executives. Furthermore, participants will develop a humble mind-set that works effectively at engaging and leading people across cultures to truly engage in dialogue and appreciate different points of view. A final key benefit is the chance to experience research-based Servant-Leadership and meaningful practices for application in organizational governance. The course is primarily targeted at executives with experience in a leadership role, a desire to serve, and an interest to find evidence supporting the efficacy of Servant-Leadership within their own organizations. In light of meaningful education for future servant-leaders, the specific objectives of the course are as follows:

- To improve understanding and research-supported practices of Servant-Leadership behaviour: historically, in business, and in humble inquiry research;
- To support organizations through meaningful learning on how to develop their leaders, in service of the organization, clients and society at large;
- To develop skills in combining leading and serving for personal and organizational development;
- To achieve cross-industry learning through co-creation between learners;
- To establish cross-organizational and cross-cultural networks and gain insight from multiple perspectives.

The structure of the course offers participants a unique combination of fieldwork in a community project, connected to and stimulating humble inquiry research in their own organizations. During a dedicated workshop, participants experience the complexity of helping from a client perspective as well as a helper's perspective. In line with Schein (2013), participants learn to apply humble inquiry and overcome misunderstanding in communication. Future Servant-Leaders are also trained in the skills of active listening. This is done because one of the biggest communication problems is caused by the fact that most people do not listen in order to *understand*, but tend to listen to *reply*. During the course, participants consider and discuss what it is that prevents them from giving meaning to others and to their organizations. What precisely do they need to make the first move towards such meaningfulness? SLRP's new research-supported learning approach will serve participants to recognize and deal with complex leadership dilemmas in their organizations.

In summary, the SLRP course aims to bridge the gap between academia and practitioners. Reducing this gap will require considerable effort – and humility – on both sides of the divide. Through learning to apply meaningfulness in the gentle art of serving and leading, participants in this course can develop into experts ensuring good organizational governance, nurtured by the best available theory, applied participatory research and (community) practice.

In the evaluation reports submitted by the SLRP participants to the 2014-2015 course, the fieldwork conducted in community projects such as Enactus (www.enactusvu.nl), Big Brothers Big Sisters (www.bbbsamsterdam.nl) and De Nieuwe Poort (www.denieuwepoort.org) was considered a highly meaningful experience. The community projects made course participants aware of the true meaning of Servant-Leadership for disadvantaged people. To illustrate, Enactus was able to create a major difference for homeless people through their tourist guide training programme. The personal stories of guides who once lived on the streets are valuable for the history of Amsterdam and visiting tourists. First and foremost, there is the empowering social benefit for the ex-homeless. Through their meaningful contribution to society, people were given a chance to regain respect and human dignity. Another benefit is the economic gain of starting a social enterprise, where the guides can support the funding of the training programme with the help of their income, and thus improve their standard of living.

The following learning experience reported by Lidl executive Verboven (2015), member of SLRP's 2014-2015 class, shows how his fieldwork in an Enactus community project impacted the meaningfulness of his everyday work at Lidl. Verboven's sense of serving was triggered during his work for the project; his awareness was raised with respect to the needs of others in the Lidl company, in particular related to the *Voedselbank*, the Dutch food bank. In the past, Lidl's contribution to the Voedselbank was very much a one-way affair; food was donated to people in need, but any fruitful dialogue with this target group was absent. After his SLRP experience, Verboven decided to encourage interaction with these people in need, finding (educational) ways to address and improve their disadvantaged position in society. In his evaluation report, following his Enactus Community experience, Verboven argues: *'When I read about Enactus, when I saw the proverb and when I visited them and did the interviews, I knew that there are more opportunities than just giving something away. I'm convinced now that we should invest in small projects, together with Organizations like Enactus to look for opportunities and to involve people in need to empower themselves in a sustainable way. Isn't it more worthwhile to teach someone to do something than to do it for him? As business people we should be committed to use our expertise, time and resources to help a new generation of young leaders understand the promise of business to create. I think that it could be of added value to a leading company to see such opportunities and turn this into value: meaningful, measurable, lasting value that will have far-reaching impact in people's lives. This should be integrated in a modern, long-term CSR strategy. In doing this Community Project I learned what Servant Leadership means in a practical way. I saw some of the characteristics of Servant Leadership in practice and I reflected on how I could integrate this in my own leadership and more, how we could integrate the ideas of Enactus in our own company'* (Verboven, 2015).

In October 2015, the second group of executives embarked on their SLRP 2015-2016 journey. In line with the Vrije Universiteit Amsterdam's core value of societal responsibility, community service has become an even stronger component in the latest SLRP course. In the new academic year, SLRP executives are supported by MSc students; jointly, they undertake the community fieldwork and identify the key demands that the leadership of the community projects are facing. Through their field research, participants discover how they can be of service to the community project. Hailing from a variety of professional backgrounds, such as accountancy, consultancy, sales and retail, executives are able to give back to the community project. The synergy of both the community practice and classroom experiences results in a meaningful educational journey for future servant-leaders. In collaboration with the VU Community Service Programme, a further development of community service learning is envisaged. We hope and trust that the SLRP case will provide qualitative insights for the further and continued development of meaningful education for servant-leaders in the 21st century. For each of us, meaningful education is an opportunity to give back to society – and to let our hearts speak.

References

Dame, J. & Gedwin, J. (2013). Six Principles for Developing Humility as a Leader. *Harvard Business Review*. September 9. https://hbr.org/2013/09/six-principles-for developing

Fortune 500 (2016). *100 Best Companies to Work For*, http://fortune.com/best-companies/

Hampden-Turner, C. and F. Trompenaars (2015). *Nine Visions of Capitalism. Unlocking the Meanings of Wealth Creation*. Infinite Ideas Limited.

Hampden-Turner, C. (forthcoming 2016). The Paradoxes of Leadership. Said Business School, University of Oxford, Chapter to be published in interactive e-book.

Liden, R. Wayne, S. Liao, C. and Meuser, J. (2014). Servant Leadership and serving Culture: Influence on Individual and Unit Performance. *Academy of Management Journal*. 57(5) 1434-1453.

Prime, J. & Salib, E. (2014). The Best Leaders Are Humble Leaders. *Harvard Business Review*, May 12. https://hbr.org/2014/05/the-best-leaders-are-humble-leaders/

Rynes, S., Giluk, T., & Brown, K. (2007). The very separate worlds of academic and practitioner periodicals in human resource management: implications for evidence based management. *Academy of Management Journal*, 50(5), 987-1008.

Schein, E. (2013). *Humble Inquiry. The Gentle Art of Asking Instead of Telling*. Berrett Koehler Publishers.

Trompenaars, F. & Voerman, E. (2009). *Servant-Leadership across Cultures*, Oxford, Infinite Ideas.

Verboven, E. (2015). *Evaluation Report on SLRP Community Project*, Vrije Universiteit Amsterdam.

Vrije Universiteit Amsterdam (2015), *VU Instellingsplan 2015-2020*, http://issuu.com/vuuniversity/docs/vu-instellingsboek-los/0, see also www.vu.nl/community-service

Vrije Universiteit Amsterdam (2016), Community Service Program, Good Practices, See also collaboration with UCLA Community Service Learning: http://www.uei.ucla.edu/communitylearning.htm

The Serving University: A Matter of Prospective Ethics

JOHAN WEMPE

Sense of serving: a provocative combination of words when applied to a university – to academic education and research. It raises a fundamental question: what end does a university serve? Who or what is being served, and how does one determine whether a university is really fulfilling its serving purpose?

In this article, I want to offer you two thoughts: my vision on what lies at the heart of sustainability issues and, following my ideas on sustainability, my view on the changing role of university education and research.

First, let me present my vision on the essence of the sustainability issue. Sustainability, I would argue, is ultimately about ethics, about moral choices and the way in which we make them. Sustainability is about countering fragmentation, about including the other and about future generations. Since the days of Adam Smith, we have been working on more efficient ways to organize ourselves. We continuously try to make the ways in which we manufacture, distribute and consume our products more efficient. The same holds true for the way in which we process residual materials and waste, and the way in which we organize the deployment of people and capital. Adam Smith visualized this process using the example of a pin factory. Previously, a craftsman was only able to produce a few pins a day, but thanks to specialization and the division of labour, it was now possible to produce thousands of pins in a single day. Since then, we have been dividing the complex whole of society into smaller and smaller parts whose components ensure

* Johan Wempe is professor of Business Ethics in the department of Management & Organization at Vrije Universiteit Amsterdam.

ever-greater efficiency. This way of thinking is not only limited to the production of physical products, but we also apply it in sectors such as healthcare and home care, municipality administrations and, more recently, also in education and research.

An important effect of the line of reasoning mentioned above is that we do not look beyond the borders of our own territory, our own company, our own sector or our own industry. We fail to see how we are affecting others, other organizations and parts of society. We fail to see how we are part of a bigger whole. We also tend to shift potential negative effects, the so-called negative externalities, to others, to future generations, to elsewhere in the world, to others in the chain or to others in our environment. A consequence of this fragmented view of society is that we are unable to recognize opportunities beyond the borders of our own territory – and that is what sustainability is really about. Sustainability involves uncovering social structures and grasping societal opportunities *beyond* the boundaries of the company and the sector. It is about optimum societal outcome. It is about optimum result, but also about results in which the interests of future generations and people elsewhere are taken into account. When vision is blinkered and fragmented, it is impossible to achieve such an integrated perspective either on society, its underlying structures or its major societal problems.

Curiously, ethics and ethicists have supported this fragmentation process; they have in fact even promoted it. A key term in ethics is 'responsibility'. Particularly in modern ethics, this concept is interpreted in a causal sense. Responsibility is qualified as accountability – people must be held accountable for the effects of their actions. The dominant line of thinking within ethics, as applied to organizations and within society as a whole, is retrospective: we assess actions afterwards, based on their effects. We are constantly looking at situations that are undesirable or unacceptable. We then determine whose action, or the lack thereof, caused this undesirable or unacceptable situation, and then hold that person accountable. This way of looking at responsibility can lead to all sorts of complex situations. It often leaves us confused, especially when there is negligence. Another tricky affair concerns the assessment of *joint* actions. Is it possible to blame people or organizations afterwards because they were unable to act together? The effect of this retrospective interpretation of responsibility is that everyone is encouraged to cover their own backs against possible errors. In the end, this leads to a situation where everybody in an organization tries to avoid taking risks: how can you avoid being held accountable for potential unwanted effects after the event? This risk-averse attitude has major implications for the willingness to accept responsibilities beyond the strictly defined tasks of a manager, a company or a sector. It leads to a legalistic interpretation of responsibility.

Within enterprises, ethics is mainly translated into the legal domain, through small print in contracts and through rules and procedures with which managers and the organization itself try to free themselves of all the risks. One consequence is the need for control. Modern organizations are characterized by the growing numbers of controllers, compliance officers and supervisors they employ. Although we inherently need these functions and functionaries within our companies and organizations, the large increase in their numbers illustrates how we have put boundaries around our organizations and how we defend them.

Sustainability as a social task requires a different type of ethics: an ethics that is prospective. I have borrowed the term prospective ethics from Jean Fourastié who, in the post-war period, was closely involved in the establishment of European cooperation. For Jean Fourastié, prospective ethics meant reflecting on the choices that need to be made with respect to the goals that we, *as a society*, set, and with respect to the means that we use to achieve these goals. Fourastié distinguishes this prospective ethics from our traditional morality, which he sees as primarily driven by our instincts. Reflective choices are largely based on scientific findings, but also on values that we have chosen to adopt.

Fourastié developed his concept of 'prospective ethics' in a period dominated by faith in technology and the progress of society. For him, it was also linked to action – expressly what you do and what you do not do. According to Fourastié, it is important in making choices to include the result of scenario studies. This is also a form of causal reasoning. I tend to go a little further than Fourastié and separate the notion of prospective ethics completely from the causal meaning of responsibility. Prospective ethics is about the role that an individual, a company, a community, an organization or a public institution plays within the bigger picture; it is about the way in which society functions and the role that every individual and organization is playing in this whole. As part of society, a chain or a sector, we need social actors to understand their role within society from the perspective of a societal optimum and not merely to view the consequences of their choices in terms of their own instincts, their self-interest or their organization's interest.

Now I would like to introduce the second thought that I set out to discuss here: my perspective on the role of scientific education and research in this fragmented world, a world that leads to unsustainable solutions to social problems and which thus requires a prospective ethics. The same processes that unfold in other sectors of society and that lead to all kinds of sustainability issues may also be recognized within university research and teaching. In the following paragraphs, I shall describe four characteristics of modern academic education and research to illustrate the point.

Education as a production process

As happened in other sectors of society, academic teaching and research have made enormous improvements in terms of efficiency. Research universities and universities of applied sciences in the Netherlands have made certain commitments, namely through 'performance contracts' (termed 'prestatiecontracten' in Dutch) with the Ministry of Education and Research, to increase the annual percentage of students graduating within the nominal study period. Within the universities, this has been translated into all kinds of mechanisms that discipline students and teachers. Things start with the precise definition of the educational objectives concerned. These educational objectives are then translated into learning tracks, and the contributions are established of the different disciplines to these learning tracks and to the educational objectives. At the same time, courses are offered in an extremely efficient way. Students receive knowledge in bite-sized chunks. Examinations determine whether each student has studied the matter adequately and is able to reproduce it. Some 250 years ago, Adam Smith analysed the organization of a pin factory. A hundred years ago, Taylor described how efficiency could be realized in the production process by dividing an entire process into a series of activities and then organizing these activities as one would operate a machine. The same model is now being used to organize the learning process of students. We perceive the learning process as a product similar to the pins in Smith's pin factory. Past and recent excesses such as the diploma fraud uncovered at several universities of applied sciences were fuelled by the pressure on institutions and teachers to reduce study time. In other areas, such as the financial world, such incentives are regarded as perverse. The consequences of what may be termed the Taylorization of the education process are more radical for education itself. Reflection and the development of insights are difficult to translate into SMART goals and hard to operationalize in the industrial organization that education has now become.

Academic freedom for students

However, it is not all doom and gloom in higher education. We also see positive developments leading to a more integrated learning process and connecting the learning process to societal needs. Students, after having obtained a Bachelor's degree in a particular field, are now able to continue their education in a different Master's programme, possibly at another university or even in another European country. This offers universities the opportunity to align their Master's programmes with the university's profile and involve Master's students in research that is carried out at the university. In a sense, the term 'academic freedom' has been given a new look. At the end of the Middle Ages and in earlier modern times, academic freedom meant the right to travel across borders for study purposes and for visiting other universities. In the 17th and 18th centuries, it was customary for young men to make a 'Grand Tour' and to study at a number of European universities, a development which often led to these men leading dissolute lives at their fathers' expense. Leiden University was particularly popular for such travels in the Netherlands. Thanks to the Treaty of Bologna, signed in 1999 and facilitating the mobility of students and teachers in Europe, something of that academic freedom has been restored. However, we need to extend this academic freedom by encouraging students and researchers to cross the boundaries of their disciplines.

Specialization within research

As I noted above, the ability to connect Master's level education to a university's research work offers opportunities to connect education with pressing societal issues and thus to contribute to a sustainable society. That said, the question remains whether research actually performs that function. There are processes at work within research that are similar to those we recognize elsewhere in society. Academic research has become increasingly focused upon a smaller segment of the world – and often even only upon the theoretical world as reflected within the scientific journals. Researchers are held accountable for delivering a number of publications in scientific journals with high impact factors. The general principle here is that each article will address a small and precisely defined question within the relevant discipline and in this way will deepen theoretical understanding. Academics are granted more time for research by their university if they publish extensively, particularly in high-ranking journals. The impact of a scientific journal is primarily based on the number of citations of the journal by scientists in other journals. In this way, scientific journals have become a reality that is independent of the major issues that we face as a society. Here, we also see excesses. These are not just about plagiarism and self-plagiarism, but also about falsified data and the funding of research by third parties through which the independence of research may become compromised. Ironically, pressing societal issues that require research across the boundaries of disciplines are often lowly ranked, at least from a scientific perspective. This is partly due to their interdisciplinary character and to a lack of appropriate highly ranked journals, and it may also be due to a lack of fundamental theoretical analysis.

Societal role of academic education and research

There is an even more fundamental way in which the fragmentation of society and current developments within academic teaching and research are related. Education and research not only promote thinking along disciplinary lines, but they also function as a societal sector that has isolated itself from other sectors and minimizes its own costs by passing these costs onto other sectors whenever possible. As is the case in other sectors of society, we are dealing with a sector in which efficiency is achieved within its own domain and where, from a societal perspective, sub-optimal solutions are the result. When you ask university managers how the university contributes to society, or to a sustainable society, they will invariably stress the importance of well-trained students and good research. The fact that academic education and research are societal sectors that can actively cooperate with other sectors in promoting sustainability remains unmentioned. Here, there are enormous opportunities: opportunities that we could make better use of.

The key question is how education and research may contribute to thinking beyond the boundaries of disciplines, of sectors, supply chains, companies and organizations, and how they can identify social issues that can be analysed in cooperation with other sectors and that lead to joint efforts to find solutions. I believe that research universities are eminently equipped to make a major contribution to the sustainability of society. They can play a leading role in the cooperation between business, society and the public sector. Prospective ethics for academic institutions presupposes a well-developed understanding of its societal role. Here, I see three factors: the potential of learning through research, cooperation with society and value-driven learning.

Naturally, it is very important that students master the basic knowledge of the various disciplines developed in the past. It is also important that students discover that they are able not only to apply their knowledge and skills to solve yesterday's issues, as described in textbooks, but also to apply more generic knowledge to address the major social issues of tomorrow. To this end, it is vital that universities cooperate with business, civil society and governments. An important part of education must start with the problems of tomorrow: with problems that cannot be solved with existing knowledge and methods. It is therefore important that students discover how their research knowledge can be used to analyse and solve major social issues as they arise in practice. This means that cooperation with society is very much needed, and we should not forget that considerable knowledge is in fact available within society, especially within business, civil society and public institutions. We need to join forces for the benefit of our education and research programmes. This requires a different way of organizing education. Education for the future should be based on a good match between research and teaching, and it should be organized around the major societal questions that are recognized in the real world.

What is also important is the attitude that a university instils in its students. A sustainable society needs employees who understand the way in which systems exert coercion and who understand how system pressures may lead to abuses and situations where sustainability opportunities are missed. Students should learn how to cope with failing systems and how such systems prevent the recognition and realization of opportunities that arise beyond the boundaries of their own tasks, their business units, their organizations and their sectors. For students, a university should be a training ground where they develop the capacity to give their own values a voice. Preparing young people for tomorrow's world by combining scientific knowledge and moral talents lies at the heart of prospective ethics and should also lie at the heart of university education. I am convinced that understanding the societal role of universities will ultimately lead to a serving university.

Reference

Fourastié, J. (1966). *Essais de morale prospective.* Paris, Gonthier.

Serving Society by Making Teachers the Linchpin of Academic Education[1]

FRANK VAN DER DUIJN SCHOUTEN

A Bildung experience

It all happened in 1965, some 50 years ago. At the *Johannes Calvijn Lyceum*, a school of secondary education in the southern district of Rotterdam, we were in fourth grade and reading Homer's Odyssey. Our teacher, Mr. Haartsen, had just lit his fifth cigarette of that day when Mr. Hak, the janitor, brought him his morning coffee. I cannot tell the first names of either Mr. Haartsen or Mr. Hak; in 1965, pupils were not supposed to know first names of teachers or janitors. These were the days when smoking in the classroom and being served coffee were a teacher's undisputed rights.

Following his daily custom, Mr. Haartsen asked one of his pupils to read a pericope from the Odyssey while another pupil had to translate it. The piece to be dealt with that day concerned Odysseus' return to his homeland Ithaca, after he had wandered around for twenty years. On his return, however, Odysseus noticed that hardly anybody expected him back after so many years. His palace was filled with men soliciting the favour of Odysseus' wife Penelope. To size up the actual situation, Odysseus, disguised as a beggar, decides to mingle with the company of his wife's suitors. Hoping to obtain information about her husband, Penelope arranges a meeting with this beggar. During the preparations for this meeting, Homer writes, Penelope starts laughing out loud, without any obvious reason.

* Professor Frank van der Duijn Schouten is the former rector magnificus at Vrije Universiteit Amsterdam.

1 A shorter Dutch version of this paper was published in Reformatorisch Dagblad, a Dutch newspaper, on 14 November 2015.

My classmate Commer Overgaauw had to read the pericope, and I was given the task of translating it. When I arrived at the lines describing the moment when Penelope starts laughing, Mr. Haartsen interrupted me with the question 'Tell me, why would Penelope start to laugh right at that moment?' The question surprised me, since it seemed to have very little to do with the translation of the Greek text. My answer was uncertain and hesitant: 'A matter of nerves?'

I will never forget what happened next. Mr. Haartsen throws his cigarette in the ashtray, jumps from his dais and positions himself in front of my desk with a penetrating gaze. He brings his forefinger straight to my nose and solemnly states in a loud voice: 'Exactly!!!'

Mr. Haartsen's act exemplifies what lifts and transforms a lesson from mere education to Bildung. What made this teacher unforgettable to me is the surprising and rather intrusive way in which he connected situations that seemed to be completely unrelated. Although I am still grateful that he insisted that we learned the Odyssey's first sentences by heart, I have to confess that I used this knowledge in my later life only by trying to impress my grandchildren with a spontaneous citation. However, the Penelope experience has been relevant for me in a variety of circumstances in life, as it gave me a deeper understanding of human behaviour.

Improving the quality of education; who is in charge?

For me, it is crystal clear that the teacher is the linchpin around which the quality of education revolves. It is neither the government nor university administrators who improve quality; it is the teacher himself or herself. Hence, the golden rule in education is to leave methods and sources to be used in the classroom primarily at the discretion of the teacher. In making choices, personal preferences and an individual's own experience will always play a decisive role. When a young and unexperienced teacher would like to offer his students a better view on Augustine, he will most likely take Augustine himself – or a book about Augustine – as a starting point, while a more experienced teacher will likely start with a current issue or problem and involve Augustine at an appropriate moment during the discussion. The latter approach not only requires a higher level of scholarship and men and women of wide reading, but it also indicates that education is a profession in which growth and experience are essential.

If the teacher is the linchpin of good-quality education, we may well ask ourselves what role all these other people play who are never in front of a class but still seem to have a considerable finger in the educational pie: members of supervisory and executive boards, deans, vice-deans, inspectors, members of accreditation committees, public servants, ministers and state secretaries. For a start, it would be a blessing for education if all these people in leadership positions were to consider it their primary goal not to be an impediment to good-quality education. If this basic awareness has been raised, they will then be ready to strive for the next level of sophistication: distinguishing between good and bad ideas for innovation as launched by individual teachers (or groups of teachers) and becoming facilitators for the good ones. The highest level that administrators or supervisors can reach is to become a stimulator or initiator of good plans and initiatives. When an academic leader is rarely bothered by ideas for change or renewal stemming from inside his or her organization, he or she may have a peaceful life, but should sooner or later get that uneasy feeling that they might not be the right person in the right place. As innovation in education should not come from 'above' but from 'below', an educational institution faces a real problem when things remain quiet and silent 'down below'.

Comparing 1965 with 2015

I realize that merely stating opinions like the ones presented above will not help us change the course that a university will take. Universities belong to the most conservative institutions in our western society, which is probably one of the main reasons why they have survived for so many centuries. 'Changing a university is like moving a graveyard; you never get cooperation from the people inside' were the words once used by a colleague to utter his frustration. Nevertheless, I believe that the only way to change a university is by just doing it with the help of small initiatives launched at the 'bottom level'. I will substantiate this statement by raising and refuting a number of objections against my glorification of Mr. Haartsen.

Obvious objections against comparing Mr. Haartsen with an assistant, associate or full professor at a university in 2015 are (I) Mr. Haartsen's class contained 25 pupils, while an average university classroom has six times as many occupied seats; (II) he could fully concentrate on developing his teaching capabilities, while a university professor is supposed to excel in teaching as well as research; (III) he lived at a time when a teacher had full authority over what happened in his classroom, while a university professor operating in today's academic world has to take into account, amongst others, learning objectives, test procedures and accreditation requirements. All of these objections are real, but I am nevertheless convinced that there are opportunities to eliminate them, as I will argue below.

Personal teaching and responsible students

One of the greatest challenges in our modern university system is to find the right balance between accommodating fairly large numbers of students and simultaneously maintaining a personal relationship between teacher and student. This problem cannot be solved easily, but our hands are not fully tied. To begin with, it would certainly help if a university were to define explicitly, for each of its educational programmes, what target levels should be set in terms of the number of students that can be accommodated without sacrificing quality. In general, a study programme with an annual influx of 75 to 100 freshmen is in a perfectly good position to survive. For programmes such as Business Studies, Economics and Law, with a considerably larger influx, organized scarcity of available student places will most likely be inevitable. However, this scarcity may be beneficial: it may increase students' awareness that being accepted while others are refused admission should go hand in hand with a greater awareness of their own responsibility.

Particularly in densely populated study programmes, the balance between right of admission and a student's own responsibility for study progress seems to be lost. Many resources can be redirected towards pure teaching activities by skipping any superfluous support that goes to students who violate the golden principle of taking their own responsibility. To illustrate the point: achievement levels for the Binding Study Advice (BSA) can be set at 90% of the programme's study credits while still allowing the dean to make justifiable exceptions. Another option is to reorganize the knowledge transfer process in such a way that students realize that coming to class unprepared is a complete waste of time, for themselves as well as for the teacher. Finally, the essence of class meetings should be shifted from the transfer of factual knowledge to learning through discussion.

Of course, the above will require a higher level of educational skills of our academics than is usually available. But why do we require an academic to follow a three-year full-time PhD programme to become a qualified researcher while educational training is restricted to the quite feeble levels set for the present Basic Educational Qualification?

To prepare students in better ways and to show them the essence of academic education, universities should restore their relationship with schools for secondary education. Too many freshmen still enter the university with completely incorrect perceptions of the journey they embark upon and of the responsibilities that are expected from a university student. Universities should blame themselves first and foremost, as they have deliberately turned their back on secondary education. It is rather worrying that universities, seats of learning that are apparently highly concerned about their impact, fail to realize that the impact of providing secondary schools with new learning materials explaining topics on the frontiers of scientific development is at least as large as the impact of a publication in a top scientific journal.

Reconciliation of education and research

In reconciling education and research at twenty-first century universities, we shouldn't forget that it is a blessing that we did not copy the model that was common in Eastern Europe in the previous century, where universities concentrated on teaching while learned institutions (Academies) were responsible for academic research. Our universities are in an ideal position to pass on the latest academic research findings to society; this can be done fast and efficiently through their educational programmes. It is precisely this link between academic education and research that should be borne in mind in allocating tasks to academics.

It is undeniable that over the past twenty years, the emphasis on quantitative measures for research output has become completely disproportionate. This was in fact amplified by the transfer of a substantial part of university funding to the Dutch National Research Council (NWO). We should be able to find ways to address and repair at least part of this imbalance. A substantial step in the right direction would be made when NWO accepts a joint responsibility with universities to safeguard the connection between research and education. One suggestion for a clear and convincing signal could be that NWO requires every NWO research proposal to include an explicit statement from applicants concerning the contributions to academic education and its innovation that they will make after receiving the coveted NWO grant.

Authority back to the teacher

I would like to make a strong plea for returning classroom authority to the individual teacher. In addition, the administrative burden placed on teachers should be kept at an absolute minimum. In obtaining any information from teachers, two questions should precede all others: 'Do we really need this information?' and 'Can we explain how this information will be used?' Accreditation procedures for study programmes should be a stimulus for the academics involved; these procedures should therefore concentrate on the content of the programme and be executed by national and international peers in the same discipline. Departmental meetings to discuss the progress and innovation of education and to safeguard its quality should become as common as research seminars are for the quality of research programmes.

Epilogue

I fully realize that mere nostalgia for the days of Mr. Haartsen will not help us move forward. Smoking in the classroom is strictly forbidden today, and quite rightly so. However, I see ample opportunities for 2016 to refocus our full and dedicated attention for quality as an inalienable characteristic of academic education. This is how universities can continue to play their role of 'servant leaders' of society as they have done from their very beginning. Full attention for educational quality at a university's lowest organizational level, i.e. its study programmes, is the best possible guarantee for fulfilling its high calling. At this moment, the appointment of the right programme directors, armed with sufficient authority and means, is the most critical success factor in reaching this goal.

The Teacher as Mentor

JAMES KENNEDY

In the summer of 1984 I spent ten weeks living and working as a volunteer in one of the poorer neighborhoods of Boston. This was part of the 'Summer of Service' program that the Midwestern American college I attended (Northwestern College in Orange City Iowa) had organized. As an American I had hoped to go abroad, but after determining the best fit they sent me to inner-city America, to an organization called Christians for Urban Justice. I had to raise my own financial support (I think it was about $ 900) by writing letters to possible donors, an ultimately successful venture. I accordingly spent the summer working in a homeless shelter by talking with its residents or helping them on errands and, on the side, working with a good government group by helping launch a project to reform the notoriously corrupt Massachusetts legislature. I don't know whether I had any impact at all, but it proved a seminal experience for me: I developed a greater appreciation for the city and its diversity, a greater sensitivity to the needs of the poor, and I realized the value of giving my own time and energy to others. It was probably the most profound experience I had during my undergraduate education, and it made me alert to the social needs of the city when I transferred my studies to Georgetown University in Washington, DC.

Many years later, I would work at another Midwestern institution, Hope College (founded by Dutch settlers), where such short-term student projects were tremendously popular; hundreds of these students (out of a total body of 3,000) volunteered each year. It was a college that took this side of student life very seriously. I had the privilege of being part of a team that wrote a successful grant to the Lilly Foundation, which around the turn of this century disbursed well more than $200 million for its Program for the Theological Exploration of Vocation to around one hundred different institutions in the United States. Like many other universities and colleges, our own vocation program was much less focused on getting students to consider a career path in the church (as the title of the program might suggest) but in getting them – very basically – to think about their own personal vocation more deeply. Central to our understanding of that concept was the definition given by the writer Frederick Buechner, who said that one's divine calling 'is the place where your deep gladness and the world's deep hunger meet'. The program we developed offered students various opportunities to reflect on what they were learning and how this related to their most important ideals. By all accounts, it was a great success, and has been continued even after Lilly funding ended, which was exactly the intent of the foundation.

In both of these cases, the role of the teacher was important. It was in the first case about teachers investing in me and discussing the plight of the city. In the second, it was the teaching staff who carried out the vocational program by leading discussion groups that were designed to help students find their vocation. In both cases, the teachers' commitments were an expression of a broader ethos at these liberal arts colleges, where care for individual students and their personal development was a central and shared passion.

When I moved to the Netherlands in 2003 to become professor of history at Vrije Universiteit Amsterdam, I entered an educational world where notions of preparing students for life-long service were far from being at the center of the educational enterprise. It was not so much a heretical idea as one removed from the central rhythms of the faculty and the university. The reasons for this difference are understandable enough. I was now a professor of Vrije Universiteit Amsterdam in a research university where the scale was larger and the purpose different than in a more student-oriented undergraduate college. It also lacked a liberal arts and science philosophy of education, something which even large American universities articulate; these universities take this philosophy seriously in order to make more systematic connections between a broad education and the imperatives of both public outreach and personal development. Additionally, I think it has mattered that the two American institutions I have mentioned were church-related, affording them a clear motivation and an available language for developing a 'sense of service' that is harder – though certainly not impossible – to replicate in a more secular setting.

That I did not think the Dutch university was sufficiently committed to a sense of serving was a belief I expressed in my inaugural lecture in 2004, where I wondered to what ends, if any, the university was training its students for. A loss or absence of mission in many institutions in Dutch society was in fact something that continued to strike me in my early years here. But despite the paucity of a strongly articulated tradition of 'service,' I also began to see and appreciate that the situation in the Netherlands was less bleak than I had at first supposed. In the first place, I noticed that both of the Amsterdam universities at which I worked were filled with colleagues who really did think that teaching was important. At least in history and the fields with which I had the most contact, I met with very few of my co-workers who were dismissive of teaching and anxious to be acquitted of it so that they might go off to 'do their own work', namely, research. Quite the contrary, I was impressed with their dedication to the teaching vocation, the thoroughness in which they marked their students' work, and the time they spent with their students. Secondly, many of them possessed the sense that they had an obligation to reach out to the broader public; they often had a keen sense that they needed to be alert to the needs of the population, whether that meant talking to civic groups, helping people with their family histories, or expressing their engagement in the media. There were exceptions, of course, but I was for the most part pleasantly surprised in what I found.

Moreover, in some respects the situation has improved since I first arrived in the Netherlands. There have been serious and structural attempts by universities and by government to pay more attention to teaching and afford it more status and resources. Utrecht University, where I presently work, has undertaken significant steps on this front since the 1990s. I know from surveys run by the university administration that teaching as a vocation has risen in status in Utrecht among faculty members in the last number of years, and this is likely the case elsewhere as well. Very gradually, one might argue that although it is still roundly dismissed as unimportant, 'the student experience' has become a bit more important in the Dutch university, in which there is more room for helping students further along in their respective lives. Additionally, the increased focus in research on societal impact ('valorization') offers more opportunities for researchers to link their own work with a more articulated sense of service, possibly also in respect to their own teaching vision.

There are, I think it is plain to see, limits to these positive developments. The central premise of this publication is that specialized research remains the leading imperative at the Dutch university, standing at the pinnacle of a hierarchy that reduces teaching and public outreach, as important as they are recognized to be, to relatively less significance. This hierarchy, in place for so long and with many deeply embedded interests, will be extremely difficult to displace.

Personally, I am less concerned about the future of teaching at the university *an sich*; as universities compete more intensely with each other and attempt to attract and retain students they will continue to place more emphasis on the quality of instruction. Student evaluations of teaching may well also receive more weight in the future. The question, though, is what they will teach, or better said, to which end they teach in the first place. But student formation, including formation that prepares them for a life of service, remains a matter about which academicians have a deep ambivalence. We know from studies in the United States that intellectual capacities such as 'critical thinking' are supported by nearly everyone in the university as an important goal of education; however, values such as enabling students to develop their own code of ethics, or come to a better self-understanding, are supported by only about one half of academicians surveyed (see Mark William Roche, *Why Choose the Liberal Arts?*, 101-104). I see this ambivalence among quite a number of my Dutch colleagues. On the one hand, they don't particularly like the idea that they've given their students a 'value-free' education that has been unmindful of, or indifferent to, the social responsibilities of students. That would be too anemic for them. On the other hand, they would feel uncomfortable with systematic discussions of ethical formation, seeing these as something that they ought not to do, or possibly, don't know how to do. This ambivalence, too, will be difficult to overcome. Formation as a theme will not be dismissed out of hand, but giving it a robust place in the educational structures of the university is another matter altogether.

Only a sustained training of instructors and a faculty culture that encourages such initiatives will be able to make a difference, and we are far from such a place today. It is not only that research has more status than teaching; in their preparation for work at the university, people spend much more time in learning how to do research than how to teach. In this respect, the introduction of the BKO is only a very small step in enhancing the teaching abilities and commitments of university staff. Yet the creation of a lively teaching culture is not only about receiving instruction, although that is important. One reason why Dutch students report back to me about 'better' teachers in the US is, I think, that American teachers have had to win students over in general education courses not intended for majors, and work harder at communicating in a liberal arts context where students come from different backgrounds. One might conclude that a mono-disciplinary approach to education, the historically dominant model in the Netherlands, does not encourage effective teaching.

Nevertheless, here, too, a slow change is taking place. In recent years, there have been more discussions in Dutch politics about the importance of *Bildung* at the university, and that is an encouraging if still a rather uncertain sign. There is a sense, too, that this ethical formation of the student, once the task of a pre-university education and since the 1960s increasingly controversial, is now being reconsidered and re-appreciated, if only because of the sense that this kind of education is really happening anywhere. The recent development of the small-scale University Colleges, a type of institution I am learning to know as a recently-appointed dean, is one place where extensive student reflection can provide a basis for a 'purpose-driven graduate' who views his life more in vocational terms than has conventionally been the case. With our 'tutor' system, in which each student is paired with a faculty advisor who spends the time to help develop a course of study best suited to the student's interests and vision, the University College goes a long way towards a model of mentoring care for the student.

It may seem from all of the above that a small-scale, liberal arts college is the only place where the best kind of learning can take place. Yet a 2014 Gallup poll of students in the United States suggests that the key to a transformational experience in college, that is, an experience that substantially increases a sense of fulfillment later on in life, both inside and outside of work, is barely related to the size of the institution (http://www.gallup.com/poll/168848/life-college-matters-life-college.aspx). What seems to matter most, at least to American students, is that they had some kind of internship experience that linked their learning to a wider world, of which my early Boston experience might also serve as an example. At the very top of the list was the experience of a teacher that made a student excited about learning; it also made a difference to many students if there was a teacher who cared about them, or was willing to serve as their mentor.

Strikingly, though, the Gallup poll also revealed that only 14 percent of Americans who had gone to college had actually had a teacher who cared about them, excited them for learning, and encouraged their dreams. The United States may be ahead on being alert to the issues discussed above, but its students are largely untouched by the deepest ways in which a university education can make a difference. Clearly, there is still a lot to do on both sides of the pond.

The importance of this kind of student experience may be growing in the Netherlands. But developing greater embodiment of this ideal remains in its infancy. It is an articulation of an ideal that still – largely – remains foreign to the research university over here. Moreover, among its supporters a further articulation of a common cause and a common language, as well as a better sense of the practices that are required to help students to become educated human beings, still needs to be explored. It is clear that such a process requires a considerable investment on the part of the dedicated teacher, someone who is truly aware that education – and its ultimate success in the life of the student – is dependent on the teacher's willingness to care and to mentor that student.

Shaping Better Professionals: The Case of Practice Based Learning

MARIO VAN VLIET

'When I was attending your class, I wondered about the added value of the material lectured. Later on, while working in practice, I realized that your class was probably the most valuable for my daily work in terms of applying the knowledge I had gathered at University.'
– Feedback from a Master's student at Radboud University Nijmegen, where I taught Information Sciences from 2003-2011.

My scientific journey

Theory and practice, practice and theory, how they interact, how theory can influence practice (application) and how practice can alter theory (adoption), how the one enriches the other and vice versa: these were themes that intrigued me when I was contemplating which major I would pursue as a graduate student. I opted for econometrics/operations research, since I very much liked the mantra of this scientific field, namely the application of theory – mathematics – to everyday problems. My PhD research focused on the specific theoretical topic of queueing theory. In queueing theory, one researches the use of mathematics for optimizing queueing systems, for instance how to minimize waiting time in a queueing system, whereby a queueing system can manifest itself in various ways: from customers waiting in line for a counter at a grocery shop to congestion of communication packages in a telecommunication network. In other words, the applicability of queuing theory to practice is widespread and easily spotted. However, I soon discovered that the systems I was analyzing, although related to practice, were an abstraction of reality, and that it was a challenge to avoid studying situations which were theoretically intriguing, but only bore little practical relevance. This became even more obvious to me in 1990, during the completion of my PhD thesis, when I visited MIT and attended a colloquium by Richard Larson,

* Mario van Vliet is professor of Information Technology Management in the department of Knowledge, Information and Networks at Vrije Universiteit Amsterdam and member of the Executive Board of Deloitte in the Netherlands.

a professor at MIT's Operations Research Center. I was raised in a school of thinking, as he was, for which optimization of queueing systems was a mathematical endeavor, meaning that we would get our scientific satisfaction by proving a system to be mathematically optimal. Richard Larson, however, showed that there was a whole other world out there. He talked about 'social justice' in queues (Larson, 1987). His theory was that, next to the objective of actual waiting time in queues, the satisfaction of customers is also influenced by the perceived justice or injustice that they experience in comparison to other waiting customers. One striking example was a case in which customers complained loudly about lengthy luggage handling delays at an airport in Texas. The management of the airport conducted a mathematical study investigating how the delays could be reduced, and it subsequently implemented a new gate allocation system for incoming planes. The results of this new system were good; they actually decreased the delays. Customer complaints, however, kept increasing. After analysis, it turned out that the total waiting time that customers experienced was determined by two factors: (a) walking time from the airplane to the luggage carrousel, and (b) waiting time at the luggage carrousel. By implementing the new allocation order, the average total waiting time had indeed decreased, but customers with checked-in luggage compared themselves with customers who had carry-on luggage, and only 'experienced' the difference in waiting time at the carrousel in a comparison between themselves and fellow passengers who did not have to wait for checked-in luggage. After learning this, the airport decided to implement some extra delays in the walking time between the airplanes and the carrousel, thereby reducing the relative difference in total waiting time between the two types of passengers, but increasing the actual average total waiting time. With this change, the system was experienced to be more socially just – and the complaints were reduced to almost zero. The conclusion of it all was that we as mathematicians had fully focused on optimizing the theoretical 'right' objective of total waiting time, but that objectives related to social justice were probably the 'right' objectives from the customers' point of view. The colloquium by Richard Larson opened my eyes and gave me a different perspective on how to apply theoretical queueing models in practice. Here, theory was used to improve practice, but at the same time practice enriched theory.

We define *practice based learning* as learning in which practical experience is used to enrich the knowledge transfer to students, with the aim to better judge practical circumstances in which knowledge might fulfill its purpose. Practice based learning comes in various ways. Without being exclusive, the most popular forms are *action based learning, problem based learning,* and *case based learning.* In all these methods, the learning trajectory is enriched by using experience from practice, either through cooperation with enterprises and public sector organizations (action based learning), by letting students, under guidance, jointly solve real-life problems in class (problem based learning), or by the use of narrated real-life cases (case based learning). The underlying idea is that when students experience how theory can be successfully applied in practice and learn about practical do's and don'ts, they will greatly benefit from this when they use the acquired knowledge in their professional careers. The use of cases in scientific teaching, for instance in the way it is done at Harvard Business School, is a prime example of how to successfully blend theory and practice. Students prepare a real-life case and later discuss this case in class. In this way, they are faced with relevant theory brought to them by the professor, and as a class, situated around the case involved, they discuss the dilemmas that managers are faced with and the type of decisions managers have to take when applying – scientific – theories in everyday situations. My personal experience as a student in one of these classes was that I left class filled with relevant knowledge, knowledge which enriched the decisions I later took in my professional career.

Practice based learning programmes require a different set of capabilities on the part of the teaching staff involved. Teachers should bring a combination of thorough scientific knowledge and true practical experience of applying theory in practice. Instructors who are able to meet this requirement and fulfill their promise to students are for example professionals who have completed their PhD and continued a professional career in which practical experience is gained, and who later, thanks to that practical experience, perform subsequent research. University professors who have conducted applied research can also be well equipped for this task. It is of prime importance that teachers should have hands-on experience through which they let their practical experience enrich theory building. Getting 'your hands dirty' by applying theory in practice is a necessary requirement.

Shaping better professionals

It is often said that there's nothing so practical as good theory. It may also be said that there's nothing so theoretically interesting as good practice (Gaffney & Anderson, 1991). As a professor at Vrije Universiteit Amsterdam, I teach graduate students. It is my strong belief that the prime objective for which I can be held accountable is to be successful in transferring knowledge to my students. A second objective is that the knowledge resides with them in their subsequent professional careers in such a way that they can apply that acquired knowledge successfully in practice. I teach complex topics such as Information Systems and Business Process Reengineering, but I am not a full-time professor. My 'daily' job is being a member of the Deloitte NL's Executive Board. Since I obtained my PhD degree, I have spent my professional career as a management consultant, and I have dealt with numerous client engagements in which I worked on topics related to the courses I currently teach. I therefore have useful professional experience relevant to my classes. By using my practical experience, I am able to illustrate, refine and expand theory by showing 'what works', 'how it works', 'why it works' and, of course, also 'what doesn't work' in practice. And what is probably even more relevant is that, based upon my personal experience, I can offer my students the reasoning why certain approaches are more successful than others.

That being said, theory and practice can sometimes contradict each other. To illustrate: a contradiction that can be used to further enhance our understanding of applicable theory is the airport example elaborated above. Relevant research is typically started by posing relevant questions. Possible contradictions between theory and practice can foster an environment in which these relevant questions emerge amongst students and teachers. In this setting, practice based learning serves as a breeding ground for the emergence of research which addresses the topics that are relevant for society and thus, in itself, can realize breakthroughs in areas which otherwise would not have been explored.

Students should gain experience in the proper judgment of both the *applicability* and the appropriate *adoption* of theory. Moreover, it is my professional conviction that I should convey to my students not only the relevant theory, but also to show them the 'norms and values' and the business ethics which in my view are essential for them to apply in their future careers. In this way, I strongly believe that I am able to shape them into becoming better professionals. Perhaps we can then form professionals who avoid taking 'unethical' decisions, which some have mentioned as a reason for the recent financial crisis that the world has gone through. For me, this is the rationale that underlines why combining theory and practice in learning is so essential. Next to the fact that this combination benefits the learning process, I am convinced that it can also enhance research. Scientific breakthroughs which are based on practical relevance are of much greater added value to our society. I encourage universities to step up their efforts and explore additional possibilities to a larger degree than is currently done to engage in joint research with corporations and other organizations.

My sense of serving

My students value the combination of theory and practice which I apply in my classes. I see this in the manner in which they 'absorb' the content I provide, I hear it in the type of questions they ask in class, and I feel it in the way in which content discussions develop during class. I *sense* that I am able to do just that which I find so thrilling in *serving* students: to transfer knowledge to upcoming graduates and thus help them to use that knowledge in the right way. Applying relevant theory, being able to judge the practical implications and, based upon this, making profoundly sound and ethically 'right' decisions is what I want my students to achieve in their future careers. It is my firm belief that as a university professor, I can be of help in shaping my students to become better professionals. If someone were to state that *my sense of serving* is focused on realizing this belief, I would certainly agree!

References

Gaffney, J. S., & Anderson, R. C. (1991). Two-tiered scaffolding: Congruent processes of teaching and learning. In E. H. Hiebert (ed.), *Literacy for a diverse society: Perspectives. Practices & policies.* NY: Teachers College Press.

Larson, R. C. (1987). Perspectives on Queues: Social Justice and the Psychology of Queueing. *Operations Research, Vol. 35, No. 6 (Nov.-Dec., 1987), 895-905.*

Great Teachers Care

MEINDERT FLIKKEMA

'Good teaching cannot be reduced to technique; good teaching comes from the identity and integrity of the teacher.'
– Parker J. Palmer, author of 'The Courage to Teach'.

Most professors in Academia have no background in education; being in front of a classroom is often a consequence of having obtained a PhD, which is basically a measure of research competency. The teaching style of these professors is a 'telling' one. They have a tendency to tell students what *they* know, sharing their knowledge without informing their audience about the path which has led to it, its boundary conditions or its assumptions. Some of them are truly brilliant scholars, but does this brilliance mean that someone is also a great academic teacher? Sometimes *yes*, they are, but in many other cases *no*, they're not for various reasons. Alternatively, many teachers who lack academic brilliance are truly great teachers. *Superb* knowing is no *conditio sine qua non* for being a great teacher. *Sufficient* knowing is!

I am firmly convinced that great teachers make great programmes, not vice versa: the quality of an education system cannot exceed the quality of its teachers (OECD, 2007). This essay therefore focuses on the question 'What makes these teachers great'? For a starter, great teachers operate at the edges of their comfort zones, and they are not intimidated to explore areas beyond their expertise. We may also ask ourselves whether great teachers are able to answer all of the questions that students raise. Perhaps they are not, but they are surely willing to engage in co-creation processes with their students and welcome their questions with professional dignity and curiosity. They master ignorance, they encourage students to tell them more, and they do not fear disagreement or not knowing. They firmly believe that questions are the drivers of learning, and therefore they are particularly interested in generating, exploring and driving questioning insights. When great teachers raise questions, they do not disqualify responses, neither verbally nor non-verbally, but they use the answers to tell their audiences stories that will resonate with them. It thrills them to search for as many perspectives as possible, and they often apply metaphors to bring ideas forward and to frame learning situations perfectly. That being said, they are equally aware of the downsides of such free floating, and therefore they often limit explorations with the help of theory, models or concepts. Without any doubt, they prefer evidence to prominence and they demonstrate an excellent sense of coherence.

Although great teachers recognize the potential of distance learning, they will never deny the importance of their students' proximity. They share the opinion that live interaction is particularly instrumental to learning. Students of great teachers are touched by their presence, their wisdom, by simple messages and empathy, not by staccato attempts to transfer knowledge with the help of slide shows. Great teachers excel in listening and mentoring. Research has clearly demonstrated that great doctors are great because they had a role model in their educational environment (Paice *et al.*, 2002), be it in school or at university. Research has also clearly shown that personal contact offers the right platform for this to occur (Wright *et al.*, 1997).

Tell them about it

A postgraduate student, Janet, once told me how Professor Ron had really moved her forward with just a few words. Operating as a consultant, Janet wanted to adopt a highly innovative approach to corporate communication, but she did not quite know how to proceed. She knew from the start that she would never become an expert, and therefore she continued to stick to the communication approaches that she was already familiar with. Her fear for negative feedback and dissatisfied customers was stronger than her courage to move forward as a communication expert. Ron did not ignore this fact and was silent for a moment. He then admitted that her dilemma was by no means easy to solve, which gave her some comfort. Afterwards, he modestly advised her to tell other people about her intention to venture into a new area. Janet acted on his advice and her public declaration of intent really moved her forward; from that moment onwards, she experienced that there was no way back.

Great teachers may make mistakes, but they are willing to admit them, which makes them human in the eyes of their learners. For these reasons, their mistakes are soon forgotten and dissolved in a collaborative, creative process. Great teachers master silence and positively reinforce each small win, as learning is hard, but they are careful not to exaggerate when they pay compliments. Similarly, they are extremely demanding and hold students accountable for their work. Any possibilities of free riding are nipped in the bud, as are exploitative and bargaining behaviours.

Great teachers de-complicate things and disentangle hypotheses from chaos; they do this without simplifying reality, but by providing it with beautiful and accessible explanations. Einstein was right: if one cannot explain something to a four-year-old, then one has not mastered the subject yet. For great teachers, rigorous preparation is of key importance, but only to be able to flow and improvise with their materials and audiences, collectively moving in an emergent direction. Great teachers always have their learners' interests at heart and thus work on things that need to be worked on be it through acts, thoughts, dance, art or music.

Designing a course while running it

Professor Jim has a special talent for reconciling the dilemma of controlling a course through 'design efforts' and providing students with the autonomy that is needed to learn, i.e. to foster 'anarchy'. Designing a course not only offers a backbone to increase efficiency, but it may also signal to students a certain degree of professionalism and 'being organized', and finally it enables a systematic approach to learning. However, when designs are too detailed, they are not helpful for dealing with the heterogeneity of student needs, difficulties and career aspirations. In a first step to address this issue, Jim contextualized the assignments that he set his students (see Figure 1) with their career aspirations as inputs. However, since academic knowledge democratizes very rapidly, Jim decided in the second year to start the course just with two things: 1. intended learnings in terms of a set of situations which students would master (from easy to complex) after completing the course, and 2. the way he would grade their work. After lecture 1, he invited his students to let him know how they would benefit from his experience from then onwards. Their feedback and his own opinion about the matter were the inputs for the next step. In this way, and with the help of advances made in ICT, he actually facilitated his students' search processes; his expertise in doing search work and conducting research became much more important than the theorizing he usually did during plenary lectures.

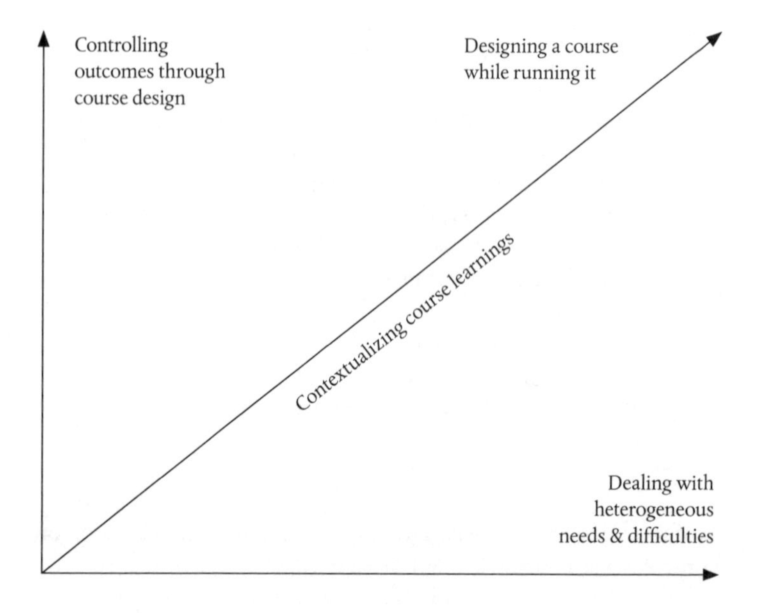

Figure 1. Reconciling control *and* anarchy *in education.*

Are great teachers providers of flashy slide shows? No, they are not; they prefer substance to superficiality, process to content, and in no way are they willing to surrender their spirit to PowerPoint. They frequently think with their hands and feel with their faces, because visualization is a powerful enabler of dialogues and a shortcut to progress. Do great teachers eloquently talk about their know-how? They sometimes do, but they make sure not to overload their audiences with everything they know and the 'borders of their knowledge' without being explicit about the difficulties they had to conquer first. They also share their doubts, and they encourage students to work with them on benefiting from these doubts in order to initiate and co-develop new research projects. This means that they live up to Von Humboldt's ideal: '*Die Einheit von Forschung und Lehre*'.

They prefer to keep a keen eye on the outside world to contribute to theory and to illustrate practical meaning, and they do not speak disdainful about colleagues who act differently.

Humboldtsches Bildungsideal

Wilhelm Von Humboldt (1767-1835) was a scholar, statesman and one of the cofounders of the University of Berlin, currently the Von Humboldt University. The University of Berlin was founded on ideas for a research-driven institution built around extraordinary privileges for an elite professorate (and also a small, elite population of students) and the principles of *Freiheit der Forschung, Freiheit der Lehre und Einheit von Forschung und Lehre*, i.e. the unity of teaching and research. Although the 19th century German model of higher education has remained an important source of inspiration until today, changes in the institutional and broader societal context have put severe pressures on practicing it.

Do great teachers provide their students with actionable feedback? They do, but they avoid using the word 'feedback' and always start with a humble inquiry (Schein, 2009) to assess their students' needs for help, to make sure that they do not help with things that students can do themselves. They may even 'read' their needs from observations. If students insist on recommendations, great teachers always provide them with at least two, to force students to select a favourite one, something which in turn makes students feel more determined to follow up these recommendations.

Do great teachers spend many hours on designing the perfect course manual? No, they do not; they go their own way and work with ingredients, not with recipes, and they lead by example. They spend many hours on studying the path that students are walking, so that they may find the right way forward. They are always on time, not only to socialize with their students but also to start on time, thus symbolizing the importance of being present and showing respect. During the kick-off sessions of new courses, great teachers pay attention to previous paths and milestones and new challenges to conquer. In this way, they show that they do not merely 'plug in' their courses, but aim to contribute to desired learning outcomes and learning trajectories. However, they also demonstrate that the path itself is a destination. They 'flip the class room' without being aware of it.

Great teachers continuously fine-tune the learnings they want to contribute to, mastering their subject or subjects and standing tall in front of the class. They gently nudge their students forward; for them, *Bildung* is not espoused theory but theory in use. They do not merely advocate its importance, but they practice it! In their work, they deepen and accelerate the learning processes of others, and they create reciprocal relationships; in this context, the word *teach* fully embraces their activities. They never lose time working with their students, because they do not hurry. They put their watches aside when they are up on the stage and invite their students to join their flow. Students' qualitative feedback subsequently fuels their own learning process and enhances self-actualization; their students' gratitude is the highest reward that can be obtained. It keeps them going. This also holds true for students who tell them about their 'un-thought known' experiences.

Stand tall, even if you don't

Peter hates giving presentations. In essence, he is a control freak and an insecure overachiever. Although he truly has important things to say, on stage he is too nervous to be able to show it. This really frustrates him, but his professor Ireen, who has grown to become an excellent presenter, noticed it without disqualifying him. She told him that she had had similar experiences in the past. Her honesty opened Peter. Ireen told the group of students to which Peter belonged that one of her mentors once told her that the ability to accept that you do not always stand tall can offer immense relief. 'Sharing your feelings and 'your place of difficulty' with your audience is certainly no weakness', she continued. 'It is a sign of courage which gets everybody immediately focused on the ball.' Upon hearing these words, Peter started to cry, because he knew that this was a moment of truth in his life. Ireen stopped talking to welcome silence. Heaven touched the ground for a moment. She finished by saying 'We are done for today', even though there was plenty of time left.

Do great teachers fight counterproductive performance management systems that are introduced by the institutions they work for? They don't do that either. They tend to focus on their students, unless a non-negotiable minimum of morality or their 'licence to operate' is at stake. They consider it their moral duty and vocation to move their students forward and to prepare them for their roles in a rapidly changing society. They also emphasize the importance of being engaged in extra-role behaviour in society, beyond regular occupations, conveying moral duties and acting accordingly themselves.

Great teachers only give grades if they have to, because assigning grades basically damages the trust-based relationship between them and their students and because it directly threatens learner cohesion – and hence learning itself. Great teachers prefer to let their students know 'silently' that their level is on par, or even better, without introducing the risk of self-complacency. They definitely prefer A-B-C rankings over 1-10 rankings (or even more detailed ones).

Great teachers approach students' motivation as a valuable resource, realizing that they need to compete with other engagements, tasks and duties for getting attention. They know their students' names, habits, preferences and difficulties, as these lay an essential foundation for understanding who their learners are. By no means do they try to benefit from students' difficulties by letting them feel that they are one down. Great teachers might be Linked-in with their students, but they are first and foremost deeply connected, without losing their distance. They offer their students containment by challenging them and by allowing them to make mistakes – but without any risk of failure, feelings of shame or loss of face. *Great teachers care!* This is what makes them well respected, remembered and honoured. Caring reconciles the dilemma of being stern or supportive. This mimics nature, when a female polar bear pushes her cub into the water for the first time. She knows this is dangerous, and she knows that the cub has not been in the water before. But she simultaneously realizes that without it, it has no chance to become a great mighty bear of the North. Equally, in human society, caring is the greatest gift and one of the greatest talents that one human being can bestow on another.

Go for question one

A colleague once told me a story about a student, let me call her Susan, and her professor, whose name was Michael. Susan was a Master's student in Business Administration who had to pass just one more exam in order to graduate. She desperately admitted that when she entered an exam room she always experienced a strong fear of failure, and even black-outs. Not even medication made her feel more comfortable. She had just this one exam opportunity left and strongly doubted whether she would go and take it. Michael carefully listened to her and advised her to give it a try, because she would never forgive herself if she did not do it. She sighed deeply and replied that she was pretty sure that she would fly into a panic as early as at question one. Michael understood, conveyed his compassion and calmly replied that she then had to go for question one and simply accept that question two and the rest would be out of reach. She could not escape from his care, went to the exam and passed it successfully.

From good to great

How does one become a great teacher? Training may help aspiring teachers move forward to some extent. To illustrate, dilemma reconciliation can be trained, which I personally experienced in workshops with experts in this area, as can active listening, coaching skills and providing students with actionable feedback. One can also learn to contextualize learnings and to switch from an atomistic learning approach (Ramsden, 1987) to a holistic approach. In an atomistic approach, parts of a system that have to be mastered are isolated while the relationships between these parts are ignored. Conversely, a holistic approach emphasizes the importance of these relationships and starts with exposing students to simple manifestations of the system. Subsequently, the level of complexity is increased step-by-step.

Still, competence development is only part of the answer. Greatness first and foremost requires a certain number of flying hours on the part of the teacher, combined with the courage to teach (Palmer, 2010) and the desire to serve. Great teachers, however, also care about themselves. Since teaching requires sweat and energy, it is important to take time for reflection and new impulses of creativity. For young teachers striving for personal growth, it is extremely important to have mentors who care and who help them to survive in institutional contexts that coercively push individuals to prioritize differently. We are all familiar with the 'publish or perish' regime in academia. In the rat race for tenure, the primary focus of young academics lies on beating the review system of top-tier journals and pursuing publication strategies with the highest possible output. This is a highly demanding job that teaches one how to convince peers. Great teachers win their audiences over by showing their true self. In a number of recent publications (Tsui, 2013; Tsui & Jia, 2013; Aguinis *et al.*, 2014) various academics held a plea for socially responsible, humanistic scholarship and a pluralistic conceptualization of scholarly impact that departs from the current win-lose and zero-sum views that lead to false trade-offs such as research *versus* practice, rigor *versus* relevance and research *versus* education. Improving the scholarly impact requires reconciling the dilemmas containing these apparent opposites.

What lies at the heart of great teacher's output is creating reciprocal student-teacher relationships through caring (see also Tsui, 2012 or Glenn, 2000). I deeply thank my own mentors for caring about me, and I promise to do everything I can to care for – and care about – my own students, irrespective of institutional contexts, both at micro and macro levels. Today's society tends to become increasingly individualized and polarized. Caring is a powerful way of uniting people, of bridging the gap between the haves and the have-nots, and of building ecosystems that ultimately govern society. I consider it a blessing that writing this message has become part of today's reality as it is unfolding itself (Jaworski & Senge, 2011) to me – and that I can add my contribution.

References

Aguinis, H., Shapiro, D. L., Antonacopoulou, E. P., & Cummings, T. G. (2014). Scholarly impact: A pluralist conceptualization. *Academy of Management Learning & Education, 13*(4), 623-639.

Glenn, E. N. (2000). Creating a caring society. *Contemporary sociology, 29*(1), 84-94.

Jaworski, J., & Senge, P. (2011). *Synchronicity: The inner path of leadership*. Berrett-Koehler Publishers.

OECD (2007). *How the world's best performing school systems come out on top*. OECD

Paice, E., Heard, S., & Moss, F. (2002). How important are role models in making good doctors? *British Medical Journal, 325*(7366), 707.

Schein, E. (2009). *Helping: How to offer, give, and receive help*. Berrett-Koehler Publishers.

Palmer, P. J. (2010). *The courage to teach: Exploring the inner landscape of a teacher's life*. John Wiley & Sons.

Ramsden, P. (1987). Improving teaching and learning in higher education: The case for a relational perspective. *Studies in Higher Education, 12*(3), 275-286.

Tsui, A. S. (2013). 2012 Presidential Address – On Compassion In Scholarship: Why Should We Care? *Academy of Management Review*, *38*(2), 167-180.

Tsui, A. S. (2013). The spirit of science and socially responsible scholarship. *Management and Organization Review*, *9*(3), 375-394.

Tsui, A. S., & Jia, L. (2013). Calling for humanistic scholarship in China. *Management and Organization Review*, *9*(1), 1-15.

Wright, S., Wong, A., & Newill, C. (1997). The impact of role models on medical students. *Journal of General Internal Medicine*, *12*(1), 53-56.

Unlocking Student Potential

SJOERD MARTENS, NISHAND SARDJOE
& TOM ZEGERS

*As soon as professors know our first names,
we are taking a large step forward.*

When we reflected on the curriculum of our Bachelor's programme, we noticed that many of the courses we followed had the same, rather boring structure: a professor in the lead and the students passively listening for ninety minutes a week, and sometimes even twice a week. Although we acknowledge that our Bachelor's trajectory was a rather generic one, with huge crowds to educate, it is a pity that only a handful of teachers managed to create truly memorable learning experiences (Fitzsimmons & Fitzsimmons, 1999) that still resonate with us because of their important take-home messages for life or insights beyond trivial theory. This small group succeeded in really moving us forward and in teaching things beyond the usual insights gained from books and papers. In their courses, these well-respected teachers would always build an overall, coherent story and start with *us* – a group of young learners, provocative and questioning relevant topics and learnings from previous classes. They were successful in creating an environment that invited students to voluntarily interact, and they inspired students to get the best out of themselves. Sometimes they would be a

* Sjoerd Martens (1992) obtained his Bachelor's degree at Vrije Universiteit Amsterdam and is currently participating in the Master's Programme Business Administration with a specialization in Strategy & Organization. Sjoerd has a keen interest in startups, business model innovation and emerging economies in Latin America.
* Nishand Sardjoe (1990) obtained his Bachelor's degree at Vrije Universiteit Amsterdam and is currently participating in the Master's Programme Business Administration with a specialization in Strategy & Organization. Nishand is an entrepreneur himself and has a keen interest in monetizing business models including sustainability.
* Tom Zegers (1989) obtained his Bachelor's and Master's degrees in Business Administration at Vrije Universiteit Amsterdam with a specialization in Strategy & Organization. Tom has a keen interest in innovation, sustainability and leadership.

bit too demanding, also for themselves, but they were always well prepared and even willing to skip things when classroom dynamics required it. The question that remains is how this approach to teaching may be implemented across every faculty, in order to increase effective teacher-student interaction and to close the social distance between teaching staff and students. To answer this question, let us start with a brief exploration of our modern academic context.

Today's universities seem to value academic publications over well-educated students who are ready for the labour market of the 21st century and who will become servants of society at large. Consequently, students are stuck with a general, suboptimal approach to teaching in which large batches of students receive lectures from a single professor to create efficiencies of scale. In ivory towers and elsewhere on many campuses, other academics benefit from these efficiencies in terms of pursuing their research projects. This approach, however, largely ignores not only the difficulties[1], dilemmas, preferences and different talents of individual students, but also their creativity. In addition, it fails to provide students with sufficient containment, which is a precondition for student engagement during lectures. Showing your doubts among hundreds of fellow students makes you vulnerable, and it may make you feel ashamed in case you fail to answer a question or come up with something in the 'heat of the moment'. Students therefore try to escape from a professor's eyes when he or she raises a question with the intention to start interaction. Only the bravest students dare to interact with professors. Sometimes this is predominantly done to show their braveness, which in fact blocks the flow of the educational process. This anonymized way of educating students – namely in large batches – has inevitably led to the standardization of education processes, major social distance between a professor and his or her students and tremendous coordination efforts. However, the latter can be reduced to a minimum if students are allowed to orchestrate their own learning process, enabled by advanced technologies. Shouldn't the university be organized as a 'job shop'[2], with physical as well as virtual education shops, as

1 Many new entrants at Vrije Universiteit Amsterdam made their start in Academia at other Dutch universities, but failed to pass the first year successfully there.

2 Job shops are typically small manufacturing systems that handle job production, that is, custom/bespoke or semi-custom/bespoke manufacturing processes such as small to medium-size customer orders or batch jobs.

opposed to faculties and departments that have been sectioned off with high walls and which merely provide mono-disciplinary education programmes? A university filled with education shops may enable customized learning trajectories that meet the preferences of individual students without offering 'easy going' routes to graduation ceremonies.

We notice that the majority of teaching staff is currently too much focused on their PowerPoint slides and Prezi presentations, and teach students in a static way – even though, admittedly, they integrate YouTube in their.ppt files – thus centralizing (and sometimes even worshipping) literature instead of centralizing the students, the communities they live in, the paths they're taking, their intended labour market position and the academic progress they aim for. Professors are generally considered to be by far the most knowledgeable in the classroom, authorities with whom you don't dare to disagree publicly, and definitely not in large groups. This is a threat to the academic freedom of students, and largely ignores their individual and collective potential, because students feel that they have to adhere to an education system without any degrees of freedom. This trend has led students to become rather passive, less creative and not very eager to develop their academic skills. Assignments and projects are predominantly seen as 'deliverables with deadlines' as opposed to great opportunities for learning and personal growth.

Within our faculty, we also notice the following phenomenon: students have multiple teachers during a single academic year. Students participate in different mandatory courses combined with a few electives. Few courses provide students with work or case groups, even though this seems to be the most appropriate setting for effective individual and team learning. The teacher who is responsible for the workgroups, however, is often not the teacher who delivers the plenary lectures, something which results in multiple approaches to teaching and multiple teachers within multiple courses, increasing the risk of confusion because of differing interpretations of course objectives or overall programme learning objectives and accents. In turn, this leads to student discouragement; without overall consistency, students will feel as if they are part of a programme without a soul, something which will also make them feel as if they were a 'number'. We would describe this education system as a *touch-and-go* system, a system with very limited possibilities of personal interaction and feedback, a key input for learning. The current situation has led to what may be qualified as 'just-passing-the-course-and-moving-on-to-the-next-one' behaviour. This type of impersonal interaction has become a pitfall in the academic education system, since it fails to focus on students' personal growth and fails to provide opportunities for creative freedom and exploring one's vocation. It also robs students of a chance to take the lead and have a say in their learning processes. The ensuing effects return with a vengeance when students have to write their Master's thesis and all of a sudden find that they *have* to take the lead!

At the moment, students still enrol at university with high expectations that (academic) freedom will follow their pre-programmed secondary school days. They hope that they will be relieved from the task – and the idea – of having to perform in a way that the teacher finds adequate. However, students are basically still hampered in finding their own path in the curriculum through their own initiatives. That's extremely disappointing. Standardized education hinders the process of strengthening the image of oneself and building confidence, simply because there is no room for a personal touch. The challenge of any faculty should be to create an environment that stimulates not only creativity, out-of-the-box thinking and an entrepreneurial mind-set, but also an 'I-care-about-society' mind-set as opposed to an egocentric 'I-don't-care' attitude. With caring about things in society we mean doing things beyond tweeting opinions and online like/dislike behaviour.

We propose to create long-lasting, meaningful relationships between teachers and students. That this can be done is shown by the Icelandic School, which launched the idea of Innovation Education (Thorsteinsson & Denton, 2003). Within the framework of their ideas, the initiators emphasize the importance of effective teacher-student interaction. They argue that educators and students should try to co-identify and co-create answers for needs and problems in our environment. The Icelandic School proposes that students and teachers cooperate and work closely in a junior peer – senior peer relationship. This will lead to teachers getting to know their students much better, and vice versa. In turn, this will result in a better understanding of specific student needs, which will then allow teachers to really help them, to enhance specific qualities and to collectively explore keen interests. In this way, teachers also act as mentors who care about their students. Gummesson (1991) mentioned the importance of emotional ties as a precondition for knowledge transfer in the relationship between teacher and student. Without strong emotional ties, knowledge transfer will surely remain an illusion.

'Erudition' and 'scholarship' are important values in Academia. This has significant consequences for the way in which academics generally prefer to teach, namely by showing their scholarship through oration. However, with this form of 'knowledge push', professors forget to listen to the interests and reasoning of students, and to feel empathy towards *their* actualities! Moreover, they may hide and not use the abstract nature of inner wisdom and consciousness (Gummesson, 1991). We mustn't forget, however, that students might also possess these qualities, qualities which are neglected in the process of one-way communication. When students are given the opportunity to choose their own mentors, they will benefit from the emotional ties mentioned above. Students' self-confidence will improve, which could in turn enhance student-teacher interaction in other disciplines, too.

The value of co-creation and its effects on the resulting quality of university teaching is also emphasized by Díaz-Méndez and Gummesson (2012). In their view, students pay tuition fees to be able to consume high-quality academic education. Students could to some extent be regarded as clients who pay a fee for the services of the faculty. They mention that value for students is not only reached through short-term student satisfaction, 'value in interaction,' but also through long-term satisfaction: 'value in use' (Lapierre, 1997). Long-term student satisfaction is shown to be determined by two crucial factors: personal development and utility for early career jobs, and the contribution that students had in a course, as compared to other students (Díaz-Méndez & Gummesson, 2012).

To offer students more academic freedom, we propose a different approach to education: a student-centred approach that entails a full focus on the needs and wishes of students, also fed by feedback from the labour market, instead of solely pushing courses that fit best with a researcher's expertise (Lin *et al.*, 2014). Students don't like to be seen as just another anonymous student, something which is all too often the case at the moment. There is no bonding with the faculty, no sense of belonging, because of the size of the institution and its 'publish or perish' regime. It is quite an anonymous life that students lead at university, and there is limited interest in one's background and future plans. Building the desired 'community of learners' should start with getting to know each other. It could even be an idea to include a personal page for every student with information about interests, extracurricular activities, part-time jobs, businesses or bucket lists, to name but a few examples.

Students should also be given the opportunity to introduce their own topics and assignments, as long as they meet the learning objectives of a course or a learning trajectory. This approach will democratize education and ask more from the teaching staff, but it is also an ideal way to deepen the student-teacher relationship and to create win-win situations. Ultimately, students could even play a role in the grading process. Reviewing the work of other students is a highly effective learning method, and with a multiple raters grading approach and low inter-rater variation, we have no serious doubts about the validity of the grading. One may ask students, for example, to qualify the work of other students in three subgroups: front runners – pack members – laggards. If it should happen that students strongly disagree with each other, then the teacher should act as the arbiter, giving feedback on the basis of his or her expertise.

The time that students spend at university is a pinnacle phase in young adolescent life, where they develop not only a greater sense of self, but also a sense of interpersonal and intellectual competence and a greater commitment to developing a meaningful philosophy of life (Astin, 1993). It is a period when students' vocations emerge, which should be supported by academic staff, technologies and appropriate learning trajectories. It is of vital importance that students get the opportunity to make their own decisions about the route they want to take when it comes to following their interests and selecting preferred research topics and approaches. The ethics shared by the academic community and the university's purpose should function as the boundaries of students' academic freedom, to avoid anarchy.

We firmly believe that this could be a priceless process in which students are motivated and encouraged to follow through on their limitless interests, preferences and ideas. Any limitations imposed by society and expectations of what a person should be doing at this point in life should be largely ignored. By encouraging students to combine personal interests like their own business or part-time job with an academic learning trajectory, student engagement will doubtlessly increase: two birds, one stone. After all, approval, validation and guidance to follow your dreams from an institution such as Vrije Universiteit Amsterdam will effectively unlock student potential to serve society. Vrije Universiteit Amsterdam should therefore rethink the adjective '*vrije*' used in its name to keep attracting students in the 21st century, but without setting aside its beloved corporate identity.

To facilitate a student-centred approach, we argue that universities in general could improve student engagement by introducing smaller groups, with a more practical approach and a labour market perspective top of mind, at least in the Master's courses. Only a small percentage of Master's students has the ambition to obtain a PhD degree, so don't be afraid that Academia will be corroded by introducing an market-oriented approach to parts of the learning trajectories. Teachers of interactive workgroups should be selected by students. Giving students the opportunity to prefer one teacher to another, because of their background, work experience, specialization, assignments or teaching style, for example, will result in the above-mentioned emotional ties (Gummesson, 1991) and their positive after-effects, especially when this is combined with longer-lasting relationships.

The relationship between a teacher and a student should be viewed as a 'senior peer to junior peer relationship', which should make courses and other physical settings for learning more engaging and encouraging than a relationship characterized by social distance (Mills & Moshavi, 1999). In this way, true human contact is promoted and the gap between research and education will disappear. This endeavour requires teaching staff representatives who perceive their jobs as a calling and a vocation rather than just a task that disturbs their research projects (Gummesson, 1991; Flikkema, 2016).

To save time for student-teacher interaction for equivocal issues and to improve the efficiency of the delivery of lectures, we also propose the introduction of E-learning, especially for the coming years, in order to adapt to today's digitizing world. It is important to mention that this would not be compatible with all course designs and content; it is constrained to lectures dealing with unequivocal subjects. Shepherd and Martz (2006) concluded that the richer the delivery of distance learning becomes, the higher the reported satisfaction and communication among students and teachers will be. Here, 'richness' is defined as 'the ability of information to change understanding within a time interval' (Daft & Lengel, 1986, p. 560). To ensure highly rich delivery with respect to distance learning, we suggest implementing fast, scalable web and mobile applications to access an education platform designed in such a way that it facilitates students to interact with other students, mentors and even outsiders such as firm representatives. These systems should provide easy online access not only to all relevant content, such as articles, presentations and even live high-quality web lectures, but also to extensive communication tools. We argue that this blended way of student learning will foster the speed of learning and its thoroughness.

To continue attracting pro-active and highly motivated students, Vrije Universiteit Amsterdam should create and facilitate an inviting and comfortable 'low-threshold' environment supportive to the desired student-teacher interaction (Mills & Moshavi, 1999). This is what we call 'organizing containment for effective learning' and being receptive for the creativity of new generations. We conclude that the research dominance in universities as well as the large numbers of new students have had its negative impact on education quality. We have addressed the problems that students are facing and considered possible solutions for these issues, which are summarized in the IST-SOLL table presented below.

Today's students are shaped according to one academic mould, predominantly in large batches. There is hardly any room for a personal touch and personal development. This calls for a change in attitude on the part of the educational institutions, because competition in higher education will definitely increase as a consequence of advanced technologies. We strongly recommend centralizing the *student*: in the education process and in Academia in general. We call upon teaching staff to see their students as junior academics, co-creators of education services and co-designers of compelling learning trajectories. We invite instructors to try and help their students address other challenges they will encounter in life and to provide them with sufficient feedback, without explicitly calling it

IST	SOLL
Education delivery in large groups. One-way communication from lecturer to students.	Smaller, interactive workgroups to promote co-creation, in combination with online lectures with rich delivery.
Student anonimity.	Students are known and seen; they experience a sense of belonging.
Multiple teachers, multiple courses.	Longer, meaningful relationships between students and teachers.
Students choose courses with a pre-assigned teacher for workgroups.	Students choose their own teacher for workgroups because of this teacher's background, specialization and work experience. A mentor role is introduced to open up dialogue and explore vocational dilemmas.
Standardized education environment with 'pre-programmed' curriculum and no personal touch.	Student-centred approach with a 'junior peer to senior peer' relationship between student and teacher, combined with individualized learning trajectories.

Table 1. Unlocking student potential: from IST to SOLL.

thus. To promote co-creation, we propose longer, meaningful relationships between students and teachers as well as smaller, more interactive workgroups. In the end, faculties should have the ambition to deliver confident, technically and socially skilled alumni, with improved self-esteem, ready for the labour market and society at large.

We hope that this essay may help Vrije Universiteit Amsterdam to rid itself from institutional pressures and the type of individualism that keeps eroding its corporate identity. Finally, we believe that only with a new servant leadership approach Vrije Universiteit Amsterdam will be able to unlock itself so that it may become able to unlock the potential of new generations of students.

References

Astin, A.W. (1993). *What matters in college?: Four critical years revisited* (Vol. 1). San Francisco: Jossey-Bass.

Daft, R.L., & Lengel, R.H. (1986). Organizational information requirements, media richness and structural design. *Management science, 32*(5), 554-571.

Díaz-Méndez, M., & Gummesson, E. (2012). Value co-creation and university teaching quality: Consequences for the European Higher Education Area (EHEA). *Journal of Service Management, 23*(4), 571-592.

Fitzsimmons, J., & Fitzsimmons, M.J. (1999). *New service development: creating memorable experiences.* Sage Publications.

Flikkema, M.J. (2016). Sense of Serving. In: Flikkema (Ed.) *Sense of Serving: Reconsidering the Role of Universities in Society Now*. Amsterdam: VU University Press.

Gummesson, E. (1991). Truths and myths in service quality. *International Journal of Service Industry Management, 2*(3), 18-23.

Lin, M.H., Chuang, T.F., & Hsu, H.P. (2014). The Relationship among Teaching Beliefs, Student-Centred Teaching Concept and the Instructional Innovation. *Journal of Service Science and Management, 7*(03), 201-210.

Lapierre, J. (1997). What does value mean in business-to-business professional services? *International Journal of Service Industry Management, 8*(5), 377-397.

Mills, P.K., & Moshavi, D.S. (1999). Professional concern: managing knowledge-based service relationships. *International Journal of Service Industry Management, 10*(1), 48-67.

Rasheed, R., & Wilson, P.R. (2014). Over Education and the Influence of Job Attributes: a Study Conducted in the City of Kochi. *Journal of Services Research, 14*(2), 145-165.

Shepherd, M.M., & Martz,W. Benjamin, Jr. (2006). Media Richness Theory and the Distance Education Environment. *The Journal of Computer Information Systems, 47*(1), 114-122.

Thorsteinsson, G., & Denton, H. (2003). The development of Innovation Education in Iceland: a pathway to modern pedagogy and potential value in the UK. *Journal of Design & Technology Education, 8*(3), 172-179.

Providing Courses in the Master's Phase is No Fast Track to Mastery

OSMAN AKSOYCAN, LOES HAFKAMP
& LAURENS WALINGA

In the past few years, the quality of higher education in the Netherlands has come under increased scrutiny, not only from students and various other stakeholders in society, but also from the Dutch accreditor of higher education NVAO. Its ratings show that a growing number of higher education programmes are qualified as subpar (Volkskrant, 2014). The organization also concluded that Dutch universities tend to devote a great deal of attention to research, thereby neglecting their role as guardians of education quality (Ritsema, 2013). Senior university management representatives always point to the increasing number of students and perverse financial incentives that universities face: their financing systems are based on the number of graduating students. This carries an inherent risk, namely the risk of relaxing requirements for students in order to safeguard overall budgets for research and education. Although these factors are undeniably part of the problem, the declining quality of higher education seems to be caused by a more complex amalgam of interrelated internal and external factors.

* Osman Aksoycan (1991) holds a Bachelor's and a Master's degree in Public Administration. He is currently involved in the Business Administration Master's programme of Vrije Universiteit Amsterdam, with a specialization in Strategy and Organization. Osman has a keen interest in innovation management, sustainability and public-private partnerships.
* Loes Hafkamp (1991) holds a Bachelor's degree in Business Administration and Philosophy and a Master's degree in Change Management. She is currently involved in the Business Administration Master's programme of Vrije Universiteit Amsterdam, with a specialization in Information and Knowledge Management. Loes has a keen interest in big data, ubiquitous computing and business ethics.
* Laurens Walinga (1987) holds a Bachelor's degree in Business Administration. He is currently involved in the Business Administration Master's programme of Vrije Universiteit Amsterdam, with a specialization in Strategy and Organization. Laurens has a keen interest in group dynamics, innovation and healthcare.

One of the factors that deserve greater attention is the teaching methods employed by university professors and other staff members. More specifically, the current teaching methods and the way in which students and teachers interact can be considered inappropriate for academic *Bildung* (Ritsema, 2013), particularly in the academic Master's phase. Currently, attempts to create challenging and memorable learning experiences are thwarted by the overriding dominance of plenary lectures, high-school-like assessments and limited opportunities to reflect on course materials and designs with teachers (Wassenaar, 2013).

Bachelor's programmes should lay the foundation for independent academic thinking and for the development of basic academic skills. In Master's programmes, students should be able to reflect on theory in a more individual and personalized way, focusing on the development of their own skills, with gentle nudges from professors when needed. Therefore, we propose to stop providing regular 3 or 6 ECTS courses in the academic Master's phase, to better prepare students for the labour market and society at large, to allow them to shape their own learning trajectories, to challenge them to find their vocation and benefit from their search and research skills as junior researchers. Preferably, this new Master's 'programme' will be developed by students themselves in a dialogue with their teachers, and it will contain three to five assignments. For every assignment, it is the students themselves who will form groups and select a mentor with whom they can discuss their needs and learning objectives. In such a programme, students will also be asked to reflect on these types of formation and matching processes in a separate assignment, so that they may learn about group dynamics, e.g. about *storming, norming and performing*. This seems particularly relevant from a labour market perspective, since team-based and project-based working is the dominant work practice in contemporary business and public environments. Finally, we need to consider the use of information and communication technology. Most people agree that today's society has become increasingly digitized, as is reflected in the increasing use of mobile devices and digital platforms. In education, be it a Bachelor's programme or a Master's programme, the use of these digital developments can lead to positive effects. We will elaborate on these positive effects on education below, when we describe our fourth proposed assignment.

We propose that universities offer their Master's students learning trajectories in which the development of skills is the key learning orientation, with the ultimate purpose of mastering problems through addressing issues and meeting challenges from practice. We propose to implement a compelling programme with five assignments to prepare students for the labour market, including universities as employers, and for serving society at large. In this essay, we shall first elaborate on these five assignments: 1. starting a Master's thesis in September, 2. conducting an internship or completing a consulting assignment, 3. writing an opinion paper on a societal issue, 4. participating in a business game and 5. writing a reflection paper about individual and team learning. Subsequently, we shall explore the type of student-teacher interaction that is needed for such a 'course-free' Master's programme. Finally, we shall explain how a virtual learning environment may support this interaction, and we shall conclude this essay by listing a number of leadership implications and presenting our final remarks.

Assignment 1: the Master thesis – start in September

Universities should place greater emphasis on developing their students' analytical and critical thinking skills instead of focusing on the mere reproduction of academic knowledge. Moreover, instead of exclusively focusing on theory and its associated applications, students should learn to *benefit* from theory and theory building; student-teacher interaction is essential in facilitating this process. Getting students more engaged in research projects is one way to increase this form of interaction. Various studies have shown that students who participate in research from an early moment onwards (at undergraduate level) show higher satisfaction and success rates throughout their studies (Hathaway, Nagda & Gregerman, 2002). We propose that even undergraduates should get involved in research projects, for example in replications of previous studies, to develop an idea of what it entails to conduct research. For the Master's curriculum, we propose that students start writing their Master's thesis immediately at the beginning of the academic year. This quick start will enable them to perform solid research in which they will collect data in small groups and have more time to reflect on their research with their fellow students and supervisors. Doing this will result in an extended writing period for each individual student and create time to write an additional article about their thesis topic or an essay for the benefit of practitioners.

Assignment 2: the internship or the consulting case

From a labour market perspective, it is very important for students to experience the practical implications of academic knowledge. Therefore, assignments should be redesigned to include a more problem-based approach, emphasizing the importance of problem solving skills and peer-to-peer collaboration as opposed to worshipping theoretical rigour. This can be done not only through partnerships with firms from the public and the private sector, but also by co-producing assignment formats, so that these meet both academic standards and professional interests. We therefore suggest to make the best possible use of these partnerships and let students complete an internship on location or write an advisory report for the companies involved. Writing such a report will contribute to mastering *evidence-based consulting and management practices* as well as to developing a better sense of the business environment and potential employers. Nevile and Adam (2003) showed that in such assignments, the practical context does not rule out an academic approach. In this way, the role of the teacher changes to a role with which conditions are created for individual and team learning as opposed to pushing theories. These kinds of partnerships, for instance with accountancy and consulting firms, may bring considerable innovation impulses to education (Den Hertog, 2000). In this respect, the path that universities should take in changing their approach to teaching can be qualified as 'open innovation', as described by Chesbrough (2011): 'open' for potential employers and for students. Finally, internships should not be seen, either by students or by instructors, as an easy opportunity to earn quick study credits or to secure education funds. In a new Master's curriculum, these internships should be seen as an opportunity to validate academic insights, to benefit from frameworks and models, to learn about evidence-based work and to conduct small-scale case studies, thus creating win-win situations for both the students and their mentors.

Assignment 3: writing an opinion paper about a societal problem while benefiting from a virtual learning environment and social media

Working in groups also serves another purpose, namely that of discussing ethical dilemmas and societal issues. Students are encouraged to find topics that they care about and subsequently to discuss these topics in small groups, facilitated by the teacher. The discussion part is aimed at deepening students' understanding of the problem at hand and to learn to problematize from various angles. This type of assignment involves an organic process; its outcome will be a written opinion document, including a thorough analysis of the identified societal issue followed by the students' opinions, or various scenarios for tackling it.

An interesting addition to discussing societal problems and challenges is presented by interaction with students from different universities, locally and abroad, to dog, share and cross-fertilize ideas. Here, several students form teams together on the condition that the dominant way in which they cooperate is by making use of IT and communication technologies. On these conditions, the teams have to deliver their written end reports. Virtual teams can offer solutions by collaborating via technologies such as Skype, Enterprise Social Media, discussion boards, blogs and other knowledge sharing tools. It should be noted, however, that working in a virtual environment is relatively hard; as is argued by Ubell (2010), it is more difficult to build trust in virtual teams. People who work in a virtual environment often fail to pick up information transmitted via non-verbal signals such as the tone of voice and facial expressions (Hakkinen & Jarvela, 2006). This means that students need to develop the skills that are necessary to improve their ability to collaborate virtually, but it also creates an opportunity to prepare students for a world where online coordination is a reality (Dineen, 2005). An opinion paper about an ethical dilemma or societal problem provides an excellent opportunity to master these skills, because of the lack of face-to-face communication and the intensive dialogue that is required. However, social problems also require face-to-face meetings, in order to experience the heat of arguments, the pain behind resistance and the energy of leadership.

This assignment should also teach students how to benefit from social media in addressing the societal issue they have selected and to source ideas from the crowd beyond the academic community. In this way, students learn to design social media strategies and to analyze big qualitative data.

Assignment 4: the business game

Modern society is becoming increasingly inundated by games, not only for fun, but also for educational purposes. We call this the *gamification* of society, and with the gamification of learning processes, firms try to benefit from the motivational power of games. Since video games and virtual worlds excel in creating engagement (McGonigal, 2011), gamification has indeed proven to be valuable for educational purposes. That being said, evidence also shows that students' understanding of actually using virtual collaboration technologies remains limited (Margaryan, Littlejohn & Vojt, 2011). For example, Abbitt's (2010) study of the use of wikis in student collaboration projects demonstrated how students were able to use the wiki to post and share information: students concluded that they were unable to actually collaborate with other students via the wiki. This implies that there is a considerable difference between the regular use of technology and virtual media on the one hand and actual collaboration with the help of advanced technologies on the other hand. Collaboration via technology requires abilities that differ from the regular use of technology and virtual media. Therefore, teachers within the university should focus on developing students' abilities to collaborate virtually (Tavcar, Zabvbi, Verlinden & Duhovniket, 2005). This can be done by introducing a business game (or a policy game, for students who prefer the public context) as part of the Master's programme. A business game, as a form of simulation-based learning, enables teachers to connect theory and practice in a more engaging and contemporary way, and it offers many possibilities for the development of collaborative skills. A business game may entail the simulation of, for instance, a certain industry including its most important stakeholders, such as competitors, customers and suppliers. Several studies have shown the advantages of adopting a business game in management education: it provides a complex and realistic yet risk-free learning environment, which is ideal for experimentation (Salas, Wildman & Piccolo, 2009; Kapp, 2012). Moreover, business games are usually affordable, easy to manage and often take student learning goals into account *ex ante* (Salas *et al.*, 2009). For example, Bascoul *et al.* (2013) showed that business games have a positive effect on students' understanding of and appreciation for management theory.

Assignment 5: reflection paper about individual and team learning

Among other things, students need effective communication skills for various purposes and target audiences, an ability to work with specific types of technology and finally effective relationship building skills. This means that they have to cover enough ground to really master a skill. During this process, constant reflection and feedback are of key importance. Students are only able to develop themselves further when they know how they are progressing and which areas need to be given greater attention. By writing a reflection paper about their individual learning, team learning and team dynamics, they devote conscious attention to their learning process within and outside a team context. Teachers will give feedback to their students and students will also give feedback to each other: a 360-approach. This information provides the basis not only for writing the reflection paper, but also for adopting and eliciting new learning goals in a personal manifesto for the near future.

A fast track to mastery: required changes in student-teacher interaction

A university and its academic staff, the teachers, should offer services to fulfill students' needs, given the learning and *Bildung* objectives of a programme. In this respect, students can be considered customers of a university. We expect and hope that student-teacher interaction will evolve in a way that will stimulate a more intense learning experience in which students become more engaged and both the student and the teacher will benefit. The 'Service Value Web' described by Chesbrough (2011) can be used to establish this form of engagement. The concept states that customers should kick-start the process of creating memorable customer experiences by discussing their needs. In our course-free Master's proposal, this will imply discussing the personal learning support that students need regarding the five assignments mentioned above, not only prior to the process, but also regularly during the process. This is the first step in creating more intensive, engaging and effective learning experiences.

In the next stage, co-creation occurs, and students show the skills they have developed so far. This will enable the teacher and the student to determine the gaps in a student's set of skills and to prioritize certain skills. Consequently, the student can formulate his or her learning objectives and ask for specific guidance in order to develop these skills. In the final stage, upon completing the *humble inquiry* (Schein, 2013), the teacher will be able to help the student develop the desired skills. Compared to the conventional way of teaching, this will result in a significantly more valuable learning experience. Its outcome will be a combination of collaborative learning, with students working together in groups to achieve a common goal, and cooperative learning, with teachers helping and guiding their students in pursuing their goals (Prince, 2004). In turn, this will likely result in new perspectives and useful mutual benefits: for students as well as instructors, for current and future teaching, and finally for current and future research, fuelled amongst other things by the various Master's theses. The win-win element is obvious.

Teachers at universities should continually strive to teach students how to make Science productive for the work environment, something which students will encounter after they have finished their education. However, as mentioned when discussing the above assignments, we believe that the Master's programmes themselves need to offer this work environment, namely by means of new teaching methods. During the past few years, the labour workforce has become more and more diverse, representing different types of employees with a multitude of cultural, generational and functional backgrounds. Due to the impact of advanced technologies and the increasing globalization of many organizations (Long & Meglich, 2013), it is expected that the majority of students will be collaborating virtually in their future workplace. Virtual collaboration involves sharing and integrating knowledge while using virtual media to communicate (Zammuto, Griffith, Majchrazak, Dougherty & Faraj, 2007). These developments lead to work environments where employees will at some point collaborate with colleagues they have never met in real life (Long & Meglich, 2013). As was argued earlier in this essay, we are of the opinion that during the assignments specific attention needs to be given to developing the students' skills with respect to these new ways of working. Table 1 below shows how the development of all important skills in a new, course-free Master's programme can be promoted.

ASSIGNMENT	SKILLS PROMOTED
Master thesis	Critical thinking Problematizing skills Literature searching skills Problem solving skills Data analysis skills Academic writing Autonomous working
Internship or consulting case	Applying theory (vs. reproducing) in practical settings Evidence based working skills Consulting skills (including pyramid reporting) Context analysis skills
Opinion paper about a societal problem	Communicating and cooperating, face to face as well as virtually Societal awareness Opinion sourcing and writing Applying theory (vs. reproducing) in practical settings Social media strategizing
Business game	Basic skills in all functional areas of the value chain: Marketing, Sales, Operations, IT, General Management, Sourcing Managing team dynamics skills
Reflection paper / personal manifesto	Processing and providing feedback Reflective practitioner skills Writing skills for development purposes

Table 1. The development of skills in a course-free Master's programme.

Conclusion

In this essay, we propose the development of a new Master's trajectory for Vrije Universiteit Amsterdam's Business Administration programme, with a key role for (five) compelling assignments. By starting with the Master's thesis, students are encouraged to reflect critically on research topics that truly interest them and on the teaching staff that might help them most effectively. Additionally, students are stimulated to think about the ways in which they want to develop their academic skills even further. The internship, the opinion paper, the business game and the reflection paper contribute to the learning trajectories of all students. Throughout the year, students will develop their academic writing skills, (virtual) communication and collaboration skills, consulting skills, team dynamics management skills, and finally their reflective practitioner skills. All these skills are developed with future workplaces top of mind, something which improves the alignment between studying and having a job in the real world. We strongly believe that through high levels of student autonomy and small teams in which students can work, academic freedom can flourish and room can be made for stimulating an entrepreneurial mind-set. Together, our five proposed assignments ensure that graduates will possess a unique set of skills that match the qualities needed in future work environments and in extra-role tasks for societal purposes. The recommendations provided in this essay are visualized below, in an overview of the current IST situation and the desired SOLL situation.

IST	SOLL
Plenary lectures with high-school-like assignments which offer a minimum of reflection on course material together with the teacher.	Stop providing courses in the regular way during the Master's programme; instead, use five assignments and individual learning trajectories.
	Start the Master's programme with conducting research in the form of writing a thesis proposal.
	Increase partnerships with businesses and organizations from the public sector and co-produce assignments. Ena-

	ble students to complete an internship at these companies, to become familiar with evidence-based management and consulting, and conduct single-firm case studies. Let students write an opinion paper about a societal problem while they cooperate virtually with other universities and source opinions from society as a whole via social media. Provide a business game to introduce students to the gamification of society and familiarize them with all business functions distinguished in the value chain. Teachers and fellow students should provide sufficient feedback to move students forward. At the same time, students should write a reflection paper to reflect on their learning, progress and dilemmas.
The current curriculum requires many coordination efforts by teachers.	Self-organizing is the preferred way of organizing the curriculum. The students are in the lead for composing the teams and the project management required in the five assignments.
Knowledge transfer is king!	Skills development is king!

Table 2. From the current IST situation to the desired SOLL situation.

References

Bascoul, G., Schmitt, J., Rasolofoarison, D., Chamberlain, L., & Lee, N. (2013). Using an experiential business game to stimulate sustainable thinking in marketing education. *Journal of marketing education*, 35(2), 168-180.

Blaskovich, J. L. (2008). Exploring the effect of distance: an experimental investigation of virtual collaboration, social loafing, and group decisions. *Journal of Information Systems*, 22(1), 27-46.

Chesbrough, H. (2011). Bringing open innovation to services. *MIT Sloan Management Review*, 52(2), 85-90.

Dineen, B.R. (2005). Teamxchange: a team project experience involving virtual teams and fluid team membership. *Journal of Management Education*, 29(4), 593-616.

Hakkinen, P. and Jarvela, S. (2006). Sharing and constructing perspectives in web-based conferencing. *Computers & Education*, 47(4), 433-47.

Hathaway, R.S., Nagda, B.A., & Gregerman, S.R. (2002). The relationship of undergraduate research participation to graduate and professional education pursuit: an empirical study. *Journal of College Student Development*, 43(5), 614-631.

Hertog, P.D. (2000). Knowledge-intensive business services as co-producers of innovation. *International Journal of Innovation Management*, 4(4), 491-528.

Kapp, K.M. (2012). *The gamification of learning and instruction: game-based methods and strategies for training and education*. San Fransisco: John Wiley & Sons.

Long, L.K., & Meglich, P.A. (2013). Preparing students to collaborate in the virtual work world. *Higher Education, Skills and Work-Based Learning*, 3(1), 6-16.

MacMillan, D. (2011,). 'Gamification': A growing business to invigorate stale websites. Retrieved 25 January, 2016, from: http://www.bloomberg.com/

McGonigal, J. (2011). *Reality Is Broken: Why Games Make Us Better and How They Can Change the World*. Penguin Press. New York, NY.

Margaryan, A., Littlejohn, A. and Vojt, G. (2011). Are digital natives a myth or reality? University students' use of digital technologies. *Computers & Technology*, 56(2), 429-40.

Neville, K., & Adam, F. (2003). Integrating theory and practice in education with business games. *Informing Science*, 6, 61-73.

Prince, M. (2004). Does active learning work? A review of the research. *Journal of Engineering Education*, 93, 223-232.

Ritsema, B. (2013). Heel begrijpelijk dat studenten kletsen en slapen tijdens college. Retrieved 10 December, 2015, retrieved from: http://www.hpdetijd.nl/

Salas, E., Wildman, J.L., & Piccolo, R.F. (2009). Using simulation-based training to enhance management education. *Academy of Management Learning & Education*, 8(4), 559-573.

Schein, E.H. (2013). *Humble inquiry: The gentle art of asking instead of telling*. Berrett Koehler Publishers.

Tavcar, J., Zabvbi, R., Verlinden, J. and Duhovnik, J. (2005). Skills for effective communication and work in global product development teams. *Journal of Engineering Design*, 16(6), 557-76.

Ubell, R. (2010). Virtual team learning, *T&D*, 65(10), 53-7.

Volkskrant (2014). Opleidingen in hoger onderwijs ondermaats. Retrieved 10 December, 2015, retrieved from: http://www.volkskrant.nl/

Wassenaar, T. (2013) 'Universitaire opleidingen stellen simpelweg veel te weinig voor'. Retrieved 10 December, 2015, retrieved from: http://www.volkskrant.nl/

Zammuto, R.F., Griffith, T.L., Majchrazak, A., Dougherty, D.J. and Faraj, S. (2007). Information technology and the changing fabric of organization. *Organization Science*, 18 (5), 749-62.

Deeply Caring about the Core

ANNE-LAURETTE LEIJSER

Many of today's students are being prepared for jobs that do not exist yet, for solving problems that we do not know are problems yet, and even for using technologies that have not been invented yet. To illustrate the point: the top 10 in-demand jobs in 2010 did not even exist in 2004 (VSNU, 2013). This makes it fair to conclude that the technological developments we have seen over the past few decades in combination with social developments and globalization have a strong impact on effective education and learning possibilities in Academia. Will the traditional classroom disappear? The answer is YES, and for good reasons.

During the past few years, many academics have come to view teaching as a service encounter in which students represent the most important stakeholders. The importance of market orientation has also been acknowledged in various university settings, as a result of which many programmes have been re-engineered (Chung & McLarney, 2000). Still, one may also have certain doubts whether students should be exclusively regarded as customers to be satisfied (cf. Martens *et al.*, 2016); *Bildung* processes may sometimes require a bit of unhappy learning, something which is probably not included in students' needs lists... Nevertheless, in this essay we shall explore ways of improving academic teaching from a service management perspective and consider how we may benefit from the latest academic insights in this domain of interest.

* Anne-Laurette Leijser (1992) obtained her Bachelor's degree in Business Administration at Vrij Universiteit Amsterdam and is currently enrolled in this university's Business Administration Master's programme, with a specialization in Strategy & Organization. She has a keen interest in innovation, leadership and international expansion.

Recent developments in the service management literature focus on the co-creation of value (Díaz-Méndez & Gummesson, 2012). This generally applicable concept implies that 'organizations do not 'provide' value to someone; instead, they actively participate in a joint process where customers also play an active role, partly through direct interaction' (Díaz-Méndez & Gummesson, 2012, p. 573). Part of the value for customers lies in receiving opportunities to interact with the service provider in the way they want to: that is, being offered degrees of freedom with respect to the place of service as well as the time and the form of the service involved – in our case education. On the basis of previous developments and recent research findings, several academics expect significant changes to be introduced in the university setting and the student-teacher interaction (cf. Desai, Hart & Richards, 2008; Díaz-Méndez & Gummesson, 2012; Dunn, 2000; Huba & Freed, 2000). One of the most illustrative examples in this case is the emergence of Massive Open Online Classes (MOOCS).

The co-creation approach mentioned above should also be applied to research projects. In this way, education and research will become much more interactive: professors will rather prefer to be doing something 'with' students, for example in research projects, than doing something 'to' students (Díaz-Méndez & Gummesson, 2012). To facilitate more interactive forms of education, small work groups with fewer students are more desirable than plenary lectures with hundreds of students. Moreover, when groups are small, instructors will be able to play a different role. The teacher, in the process of losing his or her monopoly on knowledge, will first and foremost become a mentor rather than a scientist who attempts to transfer knowledge. The transfer of knowledge is of course always particularly difficult when *tacit* knowledge is concerned (Nonaka & Takaeuchi, 1995). Benefiting from someone's tacit knowledge requires close cooperation in junior peer to senior peer relationships. Nonaka and Takeuchi (1995) call this the 'socialization strategy' for knowledge sharing. The mentor also provides opportunities for multi-media learning and stimulates self-testing. According to the experts who recently reported their findings on Vrije Universiteit Amsterdam's Education Day (Amsterdam, 12 February 2016), the latter approach appears to be a very effective learning method.

Providing education services in a university is a rather complex process; this is due to the influence of many actors and factors, such as the composition of the student population, lecturers and support staff, and also the university's economic resources, laws, student selection methods and the reputation of a university. Nevertheless, it is students and lecturers who play the most important role in the co-creation process. Lecturers aim to provide value to their students and prepare PowerPoints or Prezis to enable the value delivery, but what this process should start with is understanding students' needs and preferences (Chung & McLarney, 2000) and assessing their prior knowledge. For lecturers, it is important that these needs and dilemmas are clear and that students' initial achievement levels are known in order to move them forward and reap the rewards in return. Therefore, student-teacher interaction should be intensified. This can be done by organizing feedback sessions in small groups and by giving students the ability to influence not only the curriculum but also the organization of classes.

Students will have different learning strategies and different preferred methods of teaching. To incorporate all these wishes into a teaching method is almost impossible for a single teacher. The best way to deal with this challenge is to minimize classroom teaching or to flip the classroom when classroom teaching cannot be avoided. Nowadays, students have the possibility to complete feedback forms to evaluate a course and a teacher's performance. However, their willingness to fill out these surveys is generally low, since students do not get the feeling that doing this is effective or makes sufficient sense. By making sure that students are granted a position in which they can really craft their own (individualized) learning trajectories, they will become more engaged. '*Designing a course while running it*' (Flikkema, 2016), with input from students, would be the ultimate idea. Needless to say, this idea is by no means a plea to relax an education programme's standards and final requirements.

Another reason why the influence of students on their learning trajectories and learning methods is important relates to the increasing competition in Academia. There is a stronger need for universities to differentiate themselves. Students and teachers together can influence the unique quality of education. However, quality measurement should take place with the utmost care. The information asymmetry between teachers and students makes student satisfaction surveys in many cases an invalid measure for the performance of the lecturer. Students generally lack the technical knowledge that is necessary to evaluate a teacher's professional methods and knowledge (Díaz-Méndez & Gummesson, 2012). Nevertheless, universities should explore possibilities for designing and customizing a course to the needs of students while running it, as opposed to designing a course ex ante and in a highly detailed manner. One blueprint for a single course will never fit the needs and abilities of the entire student population.

As education is increasingly perceived as a service, stakeholder satisfaction – student satisfaction in this case – becomes vitally important. Previously, student satisfaction was seen as a collection of short-term opinions resulting from an instant evaluation of students' classroom experiences (Elliot & Healy, 2001). However, current research proposes that student satisfaction should also be focused on the long term. Here, two important issues are involved. On the one hand, there is the value of the education received from lecturers, and on the other hand we need to consider implications in the teaching-learning process (Díaz-Méndez & Gummesson, 2012). Satisfaction generated by lecturers is derived from the applicability of their work in students' future jobs and from the contribution to their personal development. Therefore, introducing alumni surveys and organizing alumni meetings are of key importance for quality management purposes. Which learnings still resonate with students? Who really made a difference? Which insights and skills continue to be highly beneficial? (e.g. Flikkema, 1995).

To prepare students in the best possible way for their future careers while bearing student satisfaction in mind, it is essential to take a closer look at students' future professions and the changing market demands. Despite the fact that our current knowledge about these future jobs is likely to fall somewhat short, predictions of changes in the labour market can be made to assist us in determining what the focus of education should be. According to Kwakman (2007), many students will be employed in Professional Service Firms (PSFS). These firms share three prominent characteristics: I. they are knowledge intensive, II. they are low capital-intensive and III. they have a highly autonomous workforce. Kwakman (2007) also states that the professional service firm of the future will be distinctive in seven areas: a unique reputation in the market, partnerships, differentiation between customer groups, customer orientation on all levels, continuous innovation in services, technology as 'driver' and a network organization. Students should develop skills to be able to work successfully in such environments, for example by focusing on skills that are needed to maintain relationships, to be entrepreneurial, customer focused and to deal with modern technology. 'The labour market' should no longer be a banned term in Academia. Universities should collect data about the early and mid-career labour market positions of their graduates, so that they may adjust and stretch learning trajectories and inform potential students about labour market perspectives and their antecedents, including behavioural and personal characteristics. This will increase transparency for potential students as well as their parents, and it will offer senior managers the opportunity to benchmark their education programmes against those offered at competing universities. Given its consequences deduced from statistical inferences, antecedent analyses will make the students more aware of important choices to be made as their learning journey progresses.

Due to the growing liberalization of trade and markets, economic systems have increasingly converged. This makes it easier for Dutch companies to enter new foreign markets and, vice versa, for foreign companies to enter the Dutch market. In fact, foreign companies are currently responsible for one out of every six new jobs in the Netherlands (Rijksoverheid, 2016). As a consequence, teams have become more cross-cultural, cross-generational and will consist of people with different nationalities (Gwee, 2008). This also requires new skills. Furthermore, it requires a much better command of English, since in many cases Dutch has ceased to be the first language spoken in the workplace. Finally, universities will be faced with more international students, particularly from Asia and from emerging economies.

Becoming an entre- or intrapreneur

Innovation through the creation of new business models is crucial for achieving economic growth at the firm level. This is why companies need individuals who are able to initiate and manage business development and innovation projects. Skills related to spotting opportunities, for instance, or acquiring finance, and skills related to networking and deciding on market strategies can be vital to a company's competitive advantage – and even its very existence. Where universities are currently more focused on *re*search, emphasizing certain managerial and entrepreneurial (search) skills would also provide students with useful tools for their future jobs. Universities can address this need to develop entrepreneurial skills in contemporary society by focusing on a broader range of activities than those conducted in a classroom setting (Rasmussen & Sørheim, 2006). Research has shown that 'educating' entrepreneurship can foster someone's entrepreneurial activities (i.e. starting new companies, finding new business areas within existing companies or running a start-up business more competently) and that this can continue to have effect throughout one's entire working career (Peterman & Kennedy, 2003). Therefore, a main emphasis should be placed on learning-by-doing initiatives and experiencing the merits of networks, for example, rather than teaching individuals in a classroom setting in a dogmatic monodisciplinary, positivistic and prescriptive way of how to become entrepreneurial[1]. This requires a shift in the learning approach from gathering knowledge and being knowledgeable to developing skills, competence and ultimately mastery (Flikkema, 2016) in order to be prepared for the fast-changing business and societal environments. Moreover, closing the gap between academic curricula and business life will give students the opportunity to become familiar with different companies and industries. This will help them in making up their minds and determining what to do after they have graduated. Moreover, it stimulates the emergence of an individual's vocation.

From rigid to flexible education

During the past few years, universities have rapidly adopted various technologies for education purposes (Georgina & Olsen, 2007). Taking Vrije Universiteit Amsterdam as an example, we note that until approximately five years ago various different software systems were used for grades, schedules and course information. All of these have now been integrated into the VUnet portal. That said, research has shown that technology itself does not necessarily create educational improvements; such improvements arise from coherent instruction and assessment enabled by advanced technologies (Goldman, Lawless, Pellegrino & Plant, 2005-2006).

1 This is like telling a young child how to ride a bike. It will answer: 'I understand', but when the child actually tries to ride the bike for the very first time, it will immediately fall to the ground for lack of practice.

These technologies enable teachers to individualize learning trajectories, and they facilitate more effective approaches to instruction and learning, including switching the context of the learning process (Lawless & Pellegrino, 2007), for example through simulation tools.

Technology will continue to change. This may provide an opportunity for a radical transformation of the entire education system, but it is also important to understand the strengths and weaknesses in terms of its actual application (Bates, 2005). Distance education, gamification, individual learning trajectories and MOOCS (tuition-free courses offered by a university to anyone with Internet access) are just a few examples of innovations that have become possible thanks to technological progress. These developments, enabled by technology, contribute to the process of co-creation, a smaller gap between academic education and professional life, and thus to greater student satisfaction and a better quality of education. It will also make education more flexible and easily accessible for students. However, it should be clear that the role of technology in education should remain supportive. Effective education requires more than a virtual space, which is also why configurations of MOOCS should not form the university of the future (Arnold, 2014; Vardi, 2012). They are the building blocks of the unfinished university of the future. Developing a 'sense of community' and a 'sense of serving', amongst other things, requires meeting each other in more traditional ways and on site: the university campus.

Education is becoming less and less dependent on location, witness the emergence of distance learning. Teachers can record their classes and put them online. Lectures will be available 24/7, and students can access them on any device (i.e. phone, laptop, tablet), anytime and anywhere. This will make the teacher's work more efficient, since he or she can record lectures whenever this is convenient. Moreover, it will prevent teachers from being bored with giving the same lectures year after year or to half-empty classrooms because the lectures are videotaped anyway. These kinds of plenary lectures can serve as background information, which will give teachers the opportunity to examine certain topics more closely and in greater detail in 'class', or to let their students talk about what they have learned and what they may have misunderstood. This will not be done in a traditional classroom setting, but in small groups discussing different subjects and with students presenting problems they have faced during the learning trajectory and during teamwork. This setting is also much more representative of real-life situations, something which will help students to adapt earlier and more quickly to the work situations they will likely face after graduating. The digital environment will help the students in preparing these meetings with their teachers, which in turn will also place a greater emphasis on 'independent learning' and taking the lead in the learning process.

Higher education in Dutch HBOS *versus Academia*

The changes discussed above justify a discussion about the differences between Dutch academic education and HBO (Dutch *Hoger Beroeps Onderwijs* or universities of applied science). From a student as well as a business perspective, it has become clear that an emphasis is needed on 'teaching'[2] skills and facilitating the mastery of contemporary problems in small group settings, both remotely and on campus. Moreover, students would like to see that *worshipping* theory, theoretical rigour and theoretical contribution is ended and that *practice* (and society at large) becomes much more important. Validating, integrating and experiencing theories combined with identifying boundary conditions should be given a first priority, with the ultimate goal of developing evidence-based working skills. To this end, adopting learning-by-doing and socialization approaches would seem to be the most appropriate way to go forward (Nonaka & Takaeuchi, 1995; Peterman & Kennedy, 2003).

Universities should be better focused on preparing students for the future, for knowledge-intensive, entrepreneurial or intrapreneurial high-tech enabled jobs with predominant search tasks. This will narrow the gap between academic education and education at HBO level, except when students are enrolled in Research Master's programmes. University graduates will also become experts in learning processes, including individual and team learning, and in orchestrating the winning performance of teams, firms and alliances. In the latter case, the development of servant leadership skills makes a substantial contribution. Education at HBO level remains focused on the development of skilled individuals who will be employed as physiotherapists, facility services managers or

2 Basically, creating the right conditions for skills development.

medical laboratory technicians. Future graduates from universities, however, have T-shaped profiles and can make a difference *in* teams and *for* teams in various complex business and societal contexts. Moreover, future academics are important 'organisms' in the ecosystems that form the backbone of society. They address imbalances, polarization and discrimination in society and take the lead in societal transformation, through writing and through organizing opposition. Again, their T-profiles allow them to do so highly effectively.

In 2025, today's eight-year-olds will have the age to go to university. The university setting and student-teacher interaction will not be the same as it is today (see Table 1 for an overview of the most important changes). Ideally, education at university will be organized on a smaller scale. Students will be learning more independently by gathering background information at home in a digital environment, on their way home and at any time they prefer; they will come to the university grounds for more interactive classes and to discuss issues for which the proximity of fellow students or more capable peers is necessary, for example for reasons of community building. The role of the teacher will be a coaching one in which he or she guides students by giving them feedback and by answering or reframing questions rather than transferring knowledge. The idea that knowledge merely needs to be transferred is, and always has been, an illusion. What is truly meaningful is co-identifying problems that need to be solved and co-creating solutions in small teams with a professor; this will lead to individual learning for *all* actors involved, to win-win situations.

Furthermore, E-learning environments are expected to play an increasingly important role in education, and students will have the possibility to craft their curriculum into individual learning trajectories, thus increasing their intrinsic motivation to study and the quality of education. Gamification is another strategy that will be useful in tailoring individual learning trajectories. By simulating real-life cases, students will be better prepared for their professional working life, and competition always triggers people to give things the best they have. Furthermore, to meet students' needs and wishes, education will be more extensively linked to business, and entrepreneurship will be fostered in many courses or via internal or external 'learning shops'. Classes will be better focused on learning-by-doing in a network context. The boundaries between universities will become less and less distinct, as will the boundaries between universities and society at large. Education digitizes and democratizes! 'Life-long learning' is no longer an unfulfilled promise, but an immediate consequence of reinvented *Bildung* processes and the attitude of the scientific staff: *they live it, and love it*. Their passion is irresistible.

When it comes to adapting to rapid changes in the environment, the current educational system is no longer tenable. Our contemporary and future society is shaped by increasing human interconnection and mobility, by the pace and depth of the evolution of human ways of life, enabled by technological innovation, and finally by the effects of globalization. In order to cope with new technologies, to solve as yet unknown problems in highly diversified, international teams and to react to the dynamics of the 21st century, it is of utmost importance for universities to focus on students developing skills to deal with these issues. Skills that are basically learned by doing – the more a person has practised, the better this person becomes – will also stand one in good stead in unfamiliar and undisclosed situations.

It takes time to learn something, even for the most talented. It was encouraging to see what Vrije Universiteit Amsterdam's recently installed *rector magnificus* professor Vinod Subramaniam wrote on a digital whiteboard at the start of Vrije Universiteit Amsterdam's Education Day (12 February 2016): 'Education is our core business'. He then emphasized that it is the only core, because 'Else we would have been a research institute and not a university' (Subramaniam, 2016). Admittedly, it is inevitable that Universities at the beginning of the 21st century also have to make money in post-graduate learning trajectories, but this is why Subramaniam rightfully added 'business' to 'education is our core'. The managerial challenge he is facing concerns motivating and supporting all academics in Vrije Universiteit Amsterdam's workforce to deeply *care* about our university's new and only *core*.

IST	SOLL
Emphasis lies on gathering knowledge.	Emphasis lies on developing skills and mastering contemporary problems in business and society.

Curriculum is set by the university, as are course structures, assignments, deliverables and exams.	Students design their own learning trajectories within boundaries set by learning objectives, a university's purpose and societal needs. Universities are therefore transparent about labour market perspectives, including behavioural and non-behavioural antecedents.
Plenary lectures with large numbers of students; no containment for rich interaction.	Small work groups in which dialogue governs the interaction between peers and more capable peers.
A teacher attempts to transfer knowledge through PowerPoint presentations.	A teacher is a 'Bilder' = coach, guide, mentor, craftsman.
Focus on lectures in a classroom setting.	Focus on E-learning environment, interactive classes, individual & team learning in various contexts, on Campus and remote.
Learning by reading articles and listening to lectures in which reductionism is key.	Learning by presenting and discussing multiple perspectives on problems, asking questions (self-tests) and playing serious games.
Research is the core business.	Education about (re)search processes and learning is the core business.
Graduates turn into alumni who have to be 'managed' in order to make money for research and to attract new students for post-graduate education.	Graduates do not leave the community of learners, but help to build an ecosystem with world citizens who care about society at large and feed the alma mater with high priority research questions and funding.

Table 1: Overview of current (IST) interaction and desired (SOLL) interaction.

References

Arnold, B.J. (2014). Gamification in education. ASBBS *Proceedings, 21*(1), 32.

Bates, A.T. (2005). *Technology, e-learning and distance education.* Routledge.

Cazden, C.B. (2001). The language of teaching and learning. *The language of teaching and learning.*

Chung, E., & McLarney, C. (2000). The classroom as a service encounter: Suggestions for value creation. *Journal of Management Education,* 24(4),484-500.

Desai, M.S., Hart, J., & Richards, T.C. (2008). E-learning: Paradigm shift in education. *Education, 129*(2), 327.

Díaz-Méndez, M., & Gummesson, E. (2012). Value co-creation and university teaching quality: Consequences for the European Higher Education Area (EHEA). *Journal of Service Management, 23*(4), 571-592.

Dunn, S.L. (2000). The virtualizing of education. *The futurist,* 34(2), 34.

Elliott, K. M. and Healy, M. A. (2001), Key factors influencing student satisfaction related to recruitment and retention, *Journal of Marketing for Higher Education,* 10(4), 1-11.

Flikkema, M.J. (1995). *Economen op de arbeidsmarkt.* Centrum voor Onderzoek van het Wetenschappelijk Onderwijs Groningen.

Flikkema, M.J. (2016). Sense of Serving. In: Flikkema, M.J. (Ed.). *Sense of Serving. Reconsidering the Role of Universities in Society Now.* Amsterdam: VU University Press.

Gartner, 2011. *Gartner says by 2015 more than 50 percent of organizations that manage innovation processes will gamify those processes.* Retrieved, 1 February 2016 from: http://www.gartner.com

Georgina, D.A., & Olson, M.R. (2008). Integration of technology in higher education: A review of faculty self-perceptions. *The Internet and Higher Education, 11*(1), 1-8.

Goldman, S.R., Lawless, K., Pellegrino, J.W., & Plants, R. (2005-2006).Technology for teaching and learning with understanding. In J.M. Cooper (Ed.), Classroom teaching skills (8th ed., pp. 185 234). Boston: Houghton Mifflin.

Gwee, M.C.E. (2008). Globalization of problem-based learning (PBL): cross cultural implications. *The Kaohsiung journal of medical sciences,* 24(3), 14-22.

Hanus, M.D., & Fox, J. (2015). Assessing the effects of gamification in the classroom: A longitudinal study on intrinsic motivation, social comparison, satisfaction, effort, and academic performance. *Computers & Education, 80,* 152-161.

Huba, M.E., & Freed, J.E. (2000). Learner centered assessment on college campuses: Shifting the focus from teaching to learning. *Community College Journal of Research and Practice,* 24(9), 759-766.

Kwakman, F.E. (2007). *The Professional Service Firm of the Future. Management Challenges for Improving Firm Performance in the New Business Landscape,* Den Haag: Academic Service.

Lawless, K.A., & Pellegrino, J.W. (2007). Professional development in integrating technology into teaching and learning: Knowns, unknowns, and ways to pursue better questions and answers. *Review of educational research,* 77(4), 575-614.

Martens, S., Sardjoe. N. & Zegers, T. (2016). Unlocking student potential. In: Flikkema, M. (ed). *Sense of Serving: Reconsidering the Role of Universities in Society Now.* Amsterdam: VU University Press.

Peterman, N.E., & Kennedy, J. (2003). Enterprise education: Influencing students' perceptions of entrepreneurship. *Entrepreneurship theory and practice*, 28(2), 129-144.

Rasmussen, E.A., & Sørheim, R. (2006). Action-based entrepreneurship education. *Technovation*, 26(2), 185-194.

Rijksoverheid, 2016. *Globalisering*. Retrieved, 10 February 2016 from: https://www.rijksoverheid.nl/onderwerpen/globalisering-en handelspolitiek/inhoud/globalisering

Vardi, M.Y. (2012). Will MOOCS destroy academia? *Commun.* ACM, 55(11), 5.

VSNU, 2013. *Toekomststrategie Nederlandse Universiteiten: visie en daadkracht voor een sterke kenniseconomie*. Retrieved, 26 December 2015 from: http://www.vsnu.nl/toekomststrategie

Yale University, 2015. *Teaching Students with Different Learning Styles and Levels of Preparation*. Retrieved, 11 February 2016 from: http://ctl.yale.edu/teaching/ideas-teaching/teachingstudents-different-learning-styles-and-levels-preparation.

Professing with the Labour Market Top of the Holistic Mind

LITA NAPITUPULU

Universities are among the least innovative educators. Apart from improvements of their ICT systems, there have been very few improvements in the past decade in terms of the quality of their 'on-site' campus services. Particularly lecturers' teaching styles have long remained the same. At Dutch universities, the improvement of teaching has generally been neglected, despite its important role in producing 'knowledge elites' that are able to meet the demands of today's labour market, particularly those set by knowledge intensive firms. Nowadays, Dutch universities devote most of their efforts and budgets to research (Advalvas, 2014).

Besides education *quality* issues, Vrije Universiteit Amsterdam – like many other universities in the Netherlands and abroad – also faces several additional challenges, if not threats. The growing number of massive open online courses (MOOCS), web-based online courses for an unlimited number of participants held by professors or other experts (Wulf *et al.*, 2014), has changed the landscape of education significantly. And the rise of MOOCS has only just begun. With the increased availability of free online courses, quality education has become cheap or almost free. Universities from all over the world have begun to broadcast their lectures to be accessible for everyone. Knowledge can now be acquired from all over the world as long as people are connected to the Internet. In the case of MOOCS, time zones and physical boundaries become

* Lita Napitupulu (Indonesia, 1983) holds a Bachelor's degree in Economics from the University of Indonesia and a Master's degree in Finance (MSc) from the Duisenberg School of Finance. She is currently enrolled in a second Master's programme at Vrije Universiteit Amsterdam, majoring in Strategy and Organization.

irrelevant, as these lectures can be replayed anytime at the viewer's convenience. Stanford University, MIT, Harvard's Coursera and Khan Academy are but a few examples of freely accessible online and high-quality education providers that manage to attract millions of students. To illustrate the point: Coursera has claimed that in September 2015 it accommodated as many as 15 million students all around the world. This shows that MOOCs play a significant role in higher education. Although the concept is still in its infancy, it may very well prove to be a threat to the market position of conventional universities. It is definitely a disruptive form of innovation in Academia.

Furthermore, competition among universities has become more fierce and more global; the list of top 50 universities in the world is no longer dominated by American and European universities. In the past few years, the rising stars in many rankings have included universities from emerging markets, such as the National University of Singapore, Nanyang Technological University, Tsinghua University, Seoul University, Peking University and Kyoto University (QS World University Ranking, 2015). This means that students from many Asian countries have plenty of high-quality universities within their own regions. It is no longer necessary to move across the globe to pursue a higher education degree from a high-ranking US or EU-based university that used to be almost exclusively located in these parts of the world.

Another major concern for Dutch universities is the tuition fees that are charged to non-EU students. Admittedly, compared to the United Kingdom and the United States, the international tuition fee in the Netherlands is relatively low. However, compared to Germany, Belgium and Scandinavian countries, Dutch tuition fees are much higher (by a wide margin). In several German *Bundesländer* (or federal states), tuition is almost free, at least compared to the 14,000 Euros charged for tuition in the Netherlands. Free tuition would be very attractive for foreign students, also considering the fact that German universities are well known for their *Gründlichkeit* (Coughlan, 2015).

All in all, MOOCs, the continuously increasing quality of Asian universities and high tuition fees are a few of the challenges confronting Vrije Universiteit Amsterdam at this moment. These create major pressures for the university to improve its market position and to distinguish itself from both local and international competitors. This is why Vrije Universiteit Amsterdam should offer unique and value-adding programmes that cannot be obtained merely through 'collecting' MOOCS or other 'offline' competitors, and the university should make its high tuition fee worth it every penny paid by its students. One way to achieve this is by improving its services as well as keeping pace with the rapidly evolving labour market and technologies that foster learning processes (Adee, 1997).

The first step that needs to be taken is to pay closer attention towards the labour market. This is a highly topical and widely discussed issue: the Dutch newspaper Trouw, for instance, reported that Dutch universities are generally unable to inform new students about their true labour market prospects, despite being urged to do so by the Dutch national students' union LSVb. Understanding the dynamics in today's job market is absolutely crucial, because it will help Vrije Universiteit Amsterdam not only to equip its students with the various skills desired by the current labour market, but also to anticipate any future market demands (Raybourn, 2015). According to Forbes Magazine, the most sought-after employees in 2015 were people who have effective communication and problem-solving skills, good information-processing skills, data analysis and IT skills, and finally effective writing and language skills on top of a main expertise, such as accounting or sales (Adams, 2014). This was confirmed in an interview we held with Adri Simamora (2015), the Dutch regional CFO of Sandoz Pharmaceuticals. He pointed out that there are three main criteria for prospective employees: they should demonstrate *long term eligibility* (a high ability to switch between tasks) in view of changes in the company business requirements, they should be *tech-savvies* who are able to swiftly adopt new technologies, and they should be *generalists* who can perform cross-departmental activities. Our current system seems to be geared towards equipping students only with a main expertise – being specialists – without paying substantial attention to these 'secondary' but nevertheless critical labour market skills.

Questioning ist

For decades, Vrije Universiteit Amsterdam's staff has applied a main teaching style without any significant changes. According to its own website, the university adopts student-centered teaching methods, but the majority of lectures are still held in large-size classrooms with limited interaction between professors and students. Most professors indeed profess while students passively listen to the 'truth' being presented to them, without being given an opportunity to approach and explore other 'truths' from different angles and perspectives. This can be typified as minimum engagement or a limited co-creation environment. From a service theory perspective, such an environment would negatively affect the perceived quality of the service and indirectly decrease customer satisfaction (Brody *et al.*, 2011). Moreover, Vrije Universiteit Amsterdam still seems to pursue an approach where course curricula, to a large extent, are developed independently, based on the lecturer's personal judgments and preferences. However, this approach has a significant drawback as it holds a risk of redundancy and a risk that the material presented during lectures may be disconnected across different courses, or even within the same study programme. Consequently, the knowledge that students acquire might become fragmented, which prevents them from developing an inherent sense that the core skills they learn and the knowledge they accumulate are actually part of a system connected to a larger entity (Senge, 1990).

Another prominent issue concerns the grading system. In many situations, this system merely focuses on students remembering what has been learnt from their professors' 'monologues'. This method has major flaws, as it degrades students' capacity – if not kills their creativity. Students' overall academic performance should not be assessed only on the basis of a three-hour exam or be defined by 40 multiple-choice questions.

An urgent need for innovation

Vrije Universiteit Amsterdam has to improve its services in order to foster the development of students' skills that match the profiles desired by employers. We argue that there are at least three elements that need to be improved: i) the teaching approach, ii) collaboration with focal industries, and iii) organizing feedback from the labour market. This will be elaborated below.

1 *Student-centred learning*

Instead of briefing students about nascent and mature theories, lecturers may flip the classroom by asking students to illustrate theories and their boundary conditions. Teachers may also foster co-research in which students learn, in teams, to collect and synthesize information and to integrate it with developing skills of inquiry, communication, critical thinking and advanced problem-solving techniques. Taught in this way, students tend to engage more actively and gain new insights compared to students taught in traditional and rather passive ways.

One of the greatest challenges involved in student-centred learning is to engage *all* students, as students have different personalities and skills levels. Some of them are talkers, while others are thinkers and observers, and not everyone is willing to share ideas. Thus, the role of the professors is highly significant here. Professors should create a welcoming atmosphere and a sense of sharing, respect and tolerance, predominantly for diversity in all imaginable dimensions. They should also create a setting that nurtures a sense of belonging, inclusiveness and mutuality in the classroom (Mann, 2001). Such a new teaching style with a student-centred focus should effectively deal with differences in preferred learning strategies as well as with differences in characters and talents.

Furthermore, students should be prepared to work in teams with diverse cultural backgrounds and different attitudes, for instance related to different generations, in addition to being able to work in a cross-functional set-up. Ranney and Deck (1995) indicated that a majority of multinational corporations increasingly use diverse team structures to meet the challenges posed by globalization and intra-organizational change. Consequently, the nature and dynamics of teams in the workplace have changed substantially; it is not unusual for people to work with colleagues hailing from different countries with different cultures. Apart from this and thanks to the overall improved quality of life and health, people generally live longer and retire later. As a result, we can distinguish four distinct and well-represented generations within today's workforce, each with its own mind-set and work habits: the conservatives, the baby boomers, generation X and generation Y. All of these groups need to work together, side by side.

Besides generational diversity and multiculturalism, graduates also face the complexities of cross-functional cooperation and hybrid careers. In contemporary society, the boundaries between jobs and between organizations are becoming less clear cut and obvious, mainly due to the increased pace of change (Ashforth, 2001). As a result, more people experience boundaryless careers:

careers that combine many different positions within multiple organizations and even within different industries (Arthur, 1994, Hall, 2002).

While teams will benefit to a high degree from cultural differences and generational gaps in the form of expanding perspectives, diversity in teamwork is also somewhat challenging and has the potential to create certain problems, such as conflicts due to differing mindsets and communication styles. By experiencing diversity during their studies and by dwelling on the topic, students will develop a capability to work effectively and efficiently with people from different cultural backgrounds, or even functional backgrounds, via a mutual understanding of each other's behaviour (Dinges, 1983). This will be useful for their future careers.

1.1 *Collaborative learning*

The conceptualization of 'collaborative learning' has become the object of many debates, but it generally refers to an instruction method in which students work together in small groups for the purpose of achieving an academic learning objective (Gokhale, 1995). Some scholars claim that an active exchange of ideas within small groups will increase interest among participants and that it will also promote critical thinking. Johnson and Johnson (1986) documented that students who work cooperatively in a team achieve higher levels of thought and retain information longer than students who work quietly as individuals.

While there is some evidence showing the benefits of collaborative learning in the classroom, as mentioned above, it is also worth mentioning that this method will also prepare students for their future workplace. Today, many organizations and corporations face novel and complicated (business) issues, and they find answers by incorporating a variety of backgrounds, ideas and personalities into the formation of project teams. In a similar style, Vrije Universiteit Amsterdam should prepare students to work effectively in teams. Graduates should be given opportunities to collaborate with others, to manage this collaboration, to allocate resources and to integrate their own skills with those of others in order to solve complex problems. Through various forms of collaboration, students will also learn how to develop a social network with their peers, a unique skill that will be beneficial to them in their efforts to increase their social capital and employability.

The 'collaborative' learning mode requires a kind of *social contract*, not only between students but also between students and professors. This contract is considered a social contract because there is implicit agreement between each member in the collaboration to jointly work upon decisions related to the tasks that every student should complete within the boundaries of institutional barriers. The contract also includes minimum performance criteria, process and product criteria, and finally expectations concerning extra-role behaviour.

1.2 *Holistic thinking*

Konorski (1967) distinguished two different approaches to understanding brain and memory: the analytical or reductionist approach and the synthetic or holistic approach. The reductionist framework explains phenomena by breaking them down into component pieces (Konorski, 1967). Reductionism is built on the premise that knowledge is made up of elementary units of experience which are grouped, related and generalized. The main goal of this approach is to transfer knowledge from the teacher to the student, in 'parts' (Macinnis, 1995). Holistic approaches, on the other hand, are based on the idea that the properties of any system in any area of study should be seen as a whole: they cannot be determined or explained exclusively by the sum of their parts.

Although the reductionist approach has several advantages, the model has come under close scrutiny, if not attack, in the past decades. Dennett (1995), in his book 'Darwin's Dangerous Ideas' introduced the term 'greedy reductionism', which refers to a state in which complexities are underestimated and theory levels are skipped in order to fasten everything securely and neatly to its foundation. Problems are narrowly defined within disciplinary boundaries, but in view of the world's growing complexity, this approach is becoming increasingly irrelevant. Consider for example, Rietje van Dam-Mieras, Vice-Rector Magnificus of Leiden University, who has expressed serious concerns about the increasingly reductionist approach to teaching adopted by universities where subjects have become increasingly specialized and issues have been reduced and sanitized. This creates the illusion of problem solving while failing to recognize the 'partial differential' approach; it means that endeavours become increasingly academic and divorced from reality.

The holistic or constructivist approach seeks to integrate pieces of knowledge or information into a cohesive and comprehensive whole (Konorski, 1967). In this approach, learning is not seen as an accumulation of facts and associations, but rather as something that is greater than the sum of its parts. As Aristotle wrote in his Metaphysics 'The whole is more than the sum of its parts', and this resonates well with the holistic model. Furthermore, in the holistic tradition, knowledge creation can be seen as a process of transformation and development (Piaget, 1970) where 'old' knowledge is altered in the process of developing new understandings. As Fosnot (1989) emphasizes: in order to make sense of new information, students need to transform and organize and connect it to their own meaning bases. Vrije Universiteit Amsterdam can help students approach complex real-world problems by adopting a holistic, multidisciplinary approach to education. This will be discussed in the next section.

1.3 *Multidisciplinary Education*

Vrije Universiteit Amsterdam should encourage its students to participate in courses from other programmes, even beyond the borders of their own faculties, in order to broaden their perspectives on business problems and societal challenges. Such a multidisciplinary approach has clear benefits with respect to students' intellectual and academic development. Lave (1988) found that multidisciplinary approaches to education activate knowledge pollination and improve students' reasoning and communication skills (Billing, 2007). Lary *et al.* (1997) documented that this learning approach enhanced problem-solving capabilities, that it indicated improvements in working in teams and that students had the feeling they had learned more about each other's discipline.

Multidisciplinary education not only helps students to grow academically, but it also helps them to have more bargaining power in the labour market. The current and future labour market does not look for people who master only one specialist skill, but rather multiple skills. As was mentioned by Sandoz' Dutch CFO Adri Simamora (interview 2015), companies are required to operate in a complex market with uncertain consumer requirements and rapidly changing technologies. This is why companies like Sandoz are looking for employees who can work and communicate with other people across traditional departmental boundaries. It is also not uncommon that an employee is required to perform basic functions that are beyond his or her original core skills. A financial analyst, for example, should also understand the basics of sales, and a salesperson should possess certain skills in the field of mathematics and statistics. All graduates are required to specialize and gather in-depth knowledge in at least one discipline, but they are also required to gather basic knowledge and skills in other disciplines. This so-called T-shaped profile allows students to tackle problems from multiple perspectives, a precondition for a holistic approach to problem solving.

1.4 *Learning from real-life cases*

The approach advocating learning via 'real-life cases' has several advantages for students. Firstly, it provides motivation for them to apply the knowledge they accumulate (Curtis, 2001). Secondly, through this method of learning, students will not only gather first-hand experience on how they may be able to use different types of knowledge to crack a problem, but they will also *know how* they can benefit from this knowledge in the future. This stimulates evidence-based practices in various disciplines. Billing (2007) argued that the transfer of skills and knowledge is fostered when general principles of reasoning are taught together with applications in varied contexts. These are valid reasons for Vrije Universiteit Amsterdam to work more closely with various focal industries and relevant partners in order to co-design their curricula as well as to provide their students with materials such as assignments or real-life cases enabled by modern technologies.

1.5 *Learning by discovery*

Learning by discovery, a teaching method where the content of what is to be learned is not given, or only to a limited extent (Ausubel, 1963), has several advantages. It confronts students with situations in which they have to independently search for knowledge that is required to achieve the goals set for the study. This encourages them to start their inquiry open-mindedly as well as to create their own schemata. By withholding relevant information, students will be challenged to activate (re)search (Taba, 1963). This method is often considered superior to others because the results of knowledge transfer are better (Hermann, 1969). Moreover, it encourages the development of problem-solving skills through the process of inquiry (Bruner, 1961). Furthermore, McDaniel and Schlager (1990) argued that students perform better if they are not guided in their attempts to find solutions (Boleratz, 1967). A professor can indeed provide students with insights and theories; however, he or she cannot tell them how to sense situations in which these insights and theories can be benefited from. Learning by discovery takes place in many ways, depending on the subjects selected and the learning objectives set for the course. For example, students can be encouraged to engage in data interpretation and self-conducted or co-created research.

From a labour market perspective, this method will help students in their future careers when they are confronted with complex problems that require an integrated set of knowledge and smart search strategies to arrive at solutions. Because much of the relevant information may be blurred or initially unavailable, this requires a series of inquiries. Here, the problem is to understand the problem. Considering the fact that learning by discovery enables students to plan their actions and to independently identify and acquire the necessary knowledge and information, these students will have a major advantage when confronted with complex problems.

1.6 *Classroom design matters*

Re-design of the classroom will affect the quality of interaction. This is also in line with findings reported by Mersha *et al.* (1992), who argue that the service-scape does in fact matter in improving the perceived quality of a service. To illustrate the point: Vrije Universiteit Amsterdam continues to adopt a traditional classroom design where a professor stands in front of the class and students can hide in the back. In contrast, McGill University has adopted a new classroom design, one where the professor stands in the middle of the class to make him or her more easily approachable while the students are seated in a circle with tables placed next to another. In addition, all students have access to a computer to support the learning process. This design has proved, according to the professors and the students concerned, to significantly activate engagement in learning activities (TedEd, 2011). Students may also benefit from new classroom designs after graduating, namely by becoming more aware of the impact of small and radical design changes on team productivity, the effectiveness of meetings, presentations, briefings and other events that take place in organizations.

1.7 *Technology-oriented and technology-enabled learning*

Educating students to become technologically adept should become one of the objectives of all programmes offered at Vrije Universiteit Amsterdam. Sandoz' CFO Adri Simamora (interview December 2015) argued that due to technological advances the business landscape is changing very fast. Leading businesses are increasingly looking for employees who are able to swiftly adopt new technologies. Furthermore, Gray (1991) argued that integrating traditional academic materials aided with technology tools will support blended and individual learning. For this reason, students should be trained to be technologically adept by integrating advanced technologies and illustrations of their use into the curriculum. This can be achieved in two different ways. Firstly, professors should include technology trends and developments into the curriculum and secondly apply advanced technologies such as serious games in their courses. In this way, they can democratize and individualize learning trajectories.

2 *University collaboration with external parties*

Most academics have always been rather allergic to being too practical. However, academic reasoning and being practical can in fact go hand in hand. Universities can and should build a strong network with external partners such as companies, NGOs and other universities. The recent calls made by NWO as well as the EU for research grants (illustrated by the GCP, ARF and KIEM research projects) that are aimed at closer collaboration between Dutch universities and companies signify the growing importance of active collaboration between the academic world and practitioners, with respect to profit as well as nonprofit organizations. This will enhance national systems of innovation.

Este and Patel (2007) stated that connecting with one another will facilitate the exchange of knowledge and result in win-win situations on all sides. Both parties should engage in joint research projects and share the costs of building a knowledge transfer network and research facilities. The university in particular can benefit from research funding and data for research, while industry partners can benefit not only from the knowledge that is generated through research in the universities, but also from the opportunity to provide input for course manuals and learning trajectories in terms of future workforce requirements. This will also enable these organizations to pre-screen potential candidates for future employment. In addition, industry partners can be involved in teaching as guest lecturers who share practical knowledge and consider the implications of academic theories, including their boundary conditions. When we consider research collaboration with NGOS, we also notice a renewed interest in working together with universities. NGOS tend to look for evidence-based practices that can be provided by universities (Aniekwe, 2012). Universities, on the other hand, will see University-NGO collaboration as a research opportunity for their staff and their students, namely in the form of research projects and internships.

Collaboration with external parties will be beneficial to students. They will be given an opportunity to conduct internships and expand their network, which will be important for early career prospects. A multidisciplinary learning trajectory would also enable students to collaborate with people from different cultural or disciplinary backgrounds. Additionally, collaboration with external parties will offer them an opportunity to work together with more senior practitioners. In addition to expanding their social network, students can use this experience to obtain valuable insights concerning life in a real company, something that can influence their adaptability when moving from an academic to a professional environment.

3 *Organize feedback from employers and graduates*

We believe that universities should demonstrate a certain 'arrogance' concerning the learning trajectories they provide. They should view their students not as clients *per se,* although the terms comes close to what students are. Universities can benefit from insights found in service management literature, but at the end of the day a university is not a service firm: it is a university. It should operate in line with public demands, its purpose, shared beliefs and values – even though this may sometimes frustrate students. Unhappy learning may be a consequence of Bildung processes providing the 'knowledge for life' that may only pay off years after the event (cf. Jos Beishuizen's contribution elsewhere in this volume). This means that collecting feedback from graduates and employers is an absolute necessity. Particularly important in this respect is qualitative feedback: statements, opinions, likes and dislikes, and feedforward – suggestions, ideas, 'no go' areas. Furthermore, feedback should be shared among academic as well as support staff in order to improve education processes and programmes. Moreover, in alumni magazines and other university publications, universities should report about the way in which they have dealt with feedback from the labour market. Studies on service management have taught us that this may make graduates even more loyal to their *alma mater* (Van Looy *et al.*, 2003, p.140). Consequently, it might make post-graduate programmes more profitable in the long run and foster life-long learning. The latter is a wish expressed by the Dutch government, albeit one that has so far remained rather utopian.

Conclusions

We may conclude that the current approach to teaching at Vrije Universiteit Amsterdam fosters the development of skills that are desired by employers only to a limited extent. This can partly be explained by the fact that feedback from the labour market is lacking. Our recommendation would be to modify the current teaching approach to a student-centred one with the labour market top of mind and future employers as strategic partners. A focus on 'learning to learn' and effective, holistic learning strategies in teams, a focus on 'know how' as opposed to 'know what' and the development of T-profiles, partly based on continuous feedback from the labour market, will likely lead to the training and development of graduates who are ready for the challenges of contemporary society and for compelling jobs in focal industries as well as in the public sector. In sum, it will result in a learning attitude for life.

References

Adams, S. (2014). *The 10 Skills Employers Most Want In 2015 Graduates*. Retrieved on December 5th 2015 from: http://www.forbes.com/sites/lizryan/2015/12/07/five-signs-your-boss-is-out-to-get-you/.

Akuni, J., Lall, P., & Stevens, D. (2012). *Academic-NGO Collaboration in International Development*. Working paper.

Athiyaman, A. (1997). 'Linking student satisfaction and service quality perceptions: the case of university education'. *European Journal of Marketing*, Vol. 31 Iss: 7, 528-540.

Benedict, L.A, Champlin, D.T, & Pence, H.E. (2013). Exploring Transmedia: The Rip-Mix-Learn Classroom. *Journal of Chemical Education*, Vol. 90 Iss: 9, 1172-1176.

Beyer, B.K. (1984). *Improving thinking skills – defining the problem*. Phi Delta Kappan. 65, 486-490.

Billing, D. (2007). Teaching for transfer of core/key skills in higher education: Cognitive skills. *Higher Education,* Vol. 53, 483-516.

Brody, R.J., Hollebeek, L.D, Juric, B., and Ilic, A. (2011). Customer Engagement: Conceptual Domain, Fundamental Propositions, and Implications for Research. *Journal of Service Research,* Vol. 14 Iss: 3, 252-271.

Brogowicz, A.A, Delene L.M & Lyth D.M (1990). A Synthesized Service Quality Model with Managerial Implications. *International Journal of Service Industry Management,* Vol. 1 Iss: 1, 27-45.

Bruner, J.S. (1961). The act of discovery. *Harvard Educational Review,* Vol. 31 Iss: 1, 21-32.

Bryson, C., & Hand, L. (2007). The role of engagement in inspiring teaching and learning, *Innovations in Education and Teaching International,* Vol. 44 Iss: 4, 342-362.

Campione, J.C., Shapiro, A.M. & Brown, A.L. (1995). Forms of transfer in a community of learners: Flexible learning and understanding, in McKeough, A., Lupart, J. & Marini, A. (eds.), *Teaching for Transfer: Fostering Generalization in Learning.* Mahwah, NJ: Lawrence Erlbaum Associates, 35-68.

Casserly, M. (2013). *The 10 Skills That Will Get You Hired In 2013.* Retrieved on December 5th 2015 from: http://www.forbes.com/sites/meghancasserly/2012/12/10/the-10-skills-that-will-get-you-a-job-in-2013/

Chesbrough, H., & Crowther, A.K. (2006). Beyond High Tech: Early Adopters of Open Innovation in Other Industries. *R&D Management,* Vol. 36 Iss: 3, 229-236.

Cheek, D. (1992). *Thinking constructively about science, technology and society education.* Albany, NY: State University of New York Press.

Coughlan S.(2015). *How Germany abolished tuition fee.* Retrieved on February 1st 2016 from: http://www.bbc.com/news/education-34132664.

Dennett (1995). Chapter 3, Universal Acid, *Darwin's Dangerous Ideas: Evolution and the Meaning of Life,* New York, NY: Simon & Schuster Paperbacks.

Dinges, N.G., & Brislin, R.W. (Eds), (1983). 'Intercultural competence', in Landis, D. *Handbook of Intercultural Training: Issues in Theory and Design,* Vol. 1, 176-202. Pergamon Press, New York.

Este, P.D., & Patel, P. (2007). University-industry linkages in the UK: what are the factors underlying the variety of interactions with industry? *Research Policy,* Vol. 36, 1295-1313.

Gokhale, A.A. (1995). Collaborative Learning Enhances Critical Thinking. *Journal of technology education,* Vol. 7 Iss: 1, 22.

Hawkins, R. (1982). *Business and the future of education.* Sacramento, CA: Sequoia Institute.

How to Manage Different Generations. Retrieved on 10th of February from: http://guides.wsj.com/management/managing-your-people/how-to-manage-different-generations/.

Konorski, J. (1967). *Integrative Activity of the Brain: an Interdisciplinary Approach.* University of Chicago Press.

Lary, M.J., Lavigne, S.E., Muma, R.D., Jones, S.E. & Hoeft, H.J. (1997). Breaking down barriers: multidisciplinary education model. *Journal of Allied Health,* Vol. 26 Iss: 2, 63-69.

MacInnis, C. (1995). Holistic and Reductionist Approaches in Special Education: Conflicts and Common Ground. *Journal of Education,* Vol.30 Iss: 11, 7-20.

McDaniel, M.A., & Schlager, M.S. (1990). Discovery learning and transfer of problem-solving skills. *Cognition and Instruction,* Vol.7 Iss: 2, 129-159.

McGill University (2011). *Teaching and Learning Experiences in Active Learning Classrooms: Highlights*. Retrieved on December 5th 2015 from: http://ed.ted.com/on/6Fzn4QJw.

Mersha, T., & Adlakha, V. (1992). 'Attributes of Service Quality: The Consumers 'Perspective', International Journal of Service Industry Management, Vol. 3 Iss: 3, 34-45.

Morley, M. *What Is the Difference Between Project Based & Non-Project Based Organizations?* Retrieved on February 10th from: http://smallbusiness.chron.com/difference-between-project-based-nonproject-based-organizations-34050.html.

Njoo, M., & Jong, T. De. (1993). Exploratory learning with a computer simulation for control theory: Learning processes and instructional support. *Journal of Research in Science Teaching*, Vol. 30, 821-844.

Phye, D. (1991). Advice and feedback during cognitive training: Effects at acquisition and delayed transfer. *Contemporary Educational Psychology*, Vol. 16, 87-94.

Piaget, J. (1970). *Structuralism*. New York: The Humanities Press.

Ranney, J., & Deck, M. (1995), Making teams work: lessons from the leaders in new product development. *Planning Review*, Vol. 23 Iss: 4, 6-13.

Raybourn, E. (2015). *Engage Learners with Transmedia Storytelling*. Retrieved on December 5th 2015 from: https://www.youtube.com/watch?v=_j-2Ct9V9cQ.

Reductionism won't save the world. Retrieved on 1st of February from: https://www.timeshighereducation.com/comment/letters/reductionism-wont-save-the-world/407132.article.

Sadler, D.R. (1989). Formative assessment and the design of instructional systems. *Instructional Science*, Vol. 18, 119-144.

Sadler, D.R. (2008), *Transforming Holistic Assessment and Grading into a Vehicle for Complex Learning*. Chapter: Assessment, Learning and Judgement in Higher Education.

Senge, P. (1990). *The fifth discipline: The art & practice of the learning organization*. NY: Doubleday/Currency.

Stahl, G., Koschmann, T., & Suthers, D. (2006). Computer-supported collaborative learning: An historical perspective. In R.K. Sawyer (Ed.), *Cambridge handbook of the learning sciences*. Cambridge University Press, 409-426.

Understanding the Methodology: QS World University Rankings. Retrieved on 4th of November 2015 from: http://www.topuniversities.com/university-rankings-articles/world-university-rankings/understanding-methodology-qs-world-university-rankings.

Van Looy, B., Gemmel, P., & Dierdonck, R. (2003). *Services management: An integrated approach*. Pearson Education.

Wicklein, R.C., & Schell, J.W. (1995), Case Studies of Multidisciplinary Approaches to Integrating Mathematics, Science and Technology Education. *Journal of Technology Education*, Vol. 6 Iss: 2, 1-6.

Wulf, J., Blohm, I., Leimeister, J.M., & Brenner, W. (2014). Massive open online courses. *Business & Information Systems Engineering*, Vol. 6 Iss: 2, 111-114.

Walvoord, B.E., & Johnson, V. (2011). *Effective Grading: A Tool for Learning and Assessment in College*.

Sense of Serving: The Extended Story

MEINDERT FLIKKEMA

'We should remind ourselves of the number of lives we touch through teaching.'
– Paul Adler (2016, 185).

Foundations for a caring society

The current situation at many Dutch universities is one of growing dissatisfaction. Accompanied by the steady reduction of funds allocated per student, the results of various perverse incentives are becoming more and more visible. Internal as well as external parties are voicing their discontent increasingly loudly, with the student occupations of the Bungehuis and Maagdenhuis in 2015 – and the subsequent eviction of protesters by special police forces – illustrating how bad things have become. What lies at the heart of today's widespread discontentedness is the fact that universities have become estranged from their roots: the *Universitas*. This was the concept that reflected the founding ideas of the first European universities, established as early as the twelfth century, and that placed a primary focus on education. Since then, the situation has changed dramatically. The network of universities has developed into a global research industry in which heavy battles are fought over the distribution of scarce resources and in which personal research output and prominence have become more important than serving society.

In their sharp and courageous analysis published in *Trouw* newspaper on 6 March 2015, Tineke Abma and Petra Verdonk observe that this process of alienation has come to pose a threat to the oneness of education and research (known as *Humboldt's educational ideal*) and that it has led to the dehumanisation of work and the working environment as well as to the erosion of education quality. The authors plead for a reassessment and a reorientation of the role played by universities in modern society. One may well wonder whether this story will have a happy ending. It probably will; after all, one cannot stop the inevitable – it is impossible to prevent summer from arriving either (Verhulst, 2015) – but it would not surprise me if the authors' call were to be met by initial lip-service: not only on the part of managers currently operating in Dutch universities, but also on the part of politicians. What is lacking at the moment is the humility, courage, vision and decisiveness needed to turn the tide and alter the process of alienation. What we see today is a symptomatic accumulation of core academic tasks. Let us consider valorisation (finding useful applications for scientific knowledge), for instance. Initially, it became an extra academic core task for universities: a serious omen. A recent addition to the academic agenda are the so-called *community services*. This follows the managerial call to action asking us to 'give back' to society. And then the penny dropped: shouldn't serving society – and offering community services – be the ultimate synthesis of everything that a *free* university stands for? In other words, shouldn't it be serving the *Universitas* (Latin: the whole, total, the universe, the world)? The ideas on which the University was originally founded are now reclaiming their position. But how do we move from here?

At our universities, we find people talking about creating a better *balance* between education and research or about incorporating teaching performance when university staff members are due for promotion, and every now and again we see newly appointed *teaching* professors. This is also somewhat reflected in the Dutch Education Ministry's recently published 2025 – Vision for Science report, but this is precisely one of the reasons why this report demonstrates a lack of vision (De Kleijn, 2014). Although *Education* is always the first element to be mentioned in ministerial abbreviations (Education, Culture and Science; Education and Research), in practice it is always forced to accept second place. It would seem as if education is 'the gammy leg that cannot be ignored' (Flikkema, 2014).

By using words such as 'creating the right *balance*', it would seem that education and research were each other's competitors and that both feed from the same manger: a manger filled with time. 'Incorporating' teaching performance will lead to students only being given what is measured, including a lot of debate about the validity and reliability of these measurements, rather than pure education quality[1]. Finally, the very appointment of *teaching* professors confirms the existence of two different worlds in our universities.

The three solutions that universities ultimately select are insufficiently effective when it comes to reconciling education and research or confirming their mutual interconnectedness, which is absent, as Hattie and Marsh (1996) show in their meta-analysis, or only incidentally present at best. On page 529 of their paper, the authors indicate that 'it is a myth that the two are inextricably intertwined'. It still is. Additional options such as curriculum revisions tend to offer little extra relief. It is not 'what is done' that matters, but 'how things are done' (De Kleijn, 2014). What matters is *Bildung*, and possibly modern *Bildung* (Huijer, 2015) or simply formation. This concerns not just some extra job to be done, for instance via individual minor programmes. The situation can be compared to that of a manager immersed in a process of change; we cannot ask of such an individual to *also* engage in change management. In a similar vein, *Bildung* is not something that can be tested and for which exams can be passed or failed. The idea alone is a terrible idea. It would be like asking Dutch gymnastics champions to explain how to execute the Cassina-Kovacs-Kolman combination for the high bar.

Of course, we have witnessed the rise of wonderful new initiatives such as University Colleges, but these can only survive in current institutional contexts if their existence remains limited to small-scale settings. Finally, writing impressive reports and organising extensive discussions about trust, as is often seen in the world of Academia lately, will only contribute very little to the actual building of trust. We shall have to dig deeper and start liberating ourselves from the socially harmful and crippling conviction that 'only numbers tell the tale'. It is this notion that leads not only to situations in which academic staff are assessed in ways that are unsuitable and inappropriate for the purpose concerned, but also to situations in which colleagues find themselves in constant competition with other colleagues. The desire for better positions is a restraining factor and a significant barrier for everyone, eve-

1 Vrije Universiteit Amsterdam scientist Dr. Lothar Kuijper very aptly commented on this development in his essay entitled *Het onmeetbare van idealen is geen reden om het meetbare tot ideaal te verheffen* (Kuijper, 2015; p. 27). Vrije Universiteit Amsterdam's Dean Karen Maex (2015, p. 44) offers a highly illustrative description of continued attempts to objectify specific goals in terms of numbers, as if these were the *servants* of desired levels of control.

rywhere, and many can be heard lamenting that things should be organised differently and that current trends should be reversed.

What is meaningful in society, a world of which universities are an inalienable part (Maex, 2015), cannot be measured (Schouw, 2015; Van Baars, 2015). Our approach to *ratio* and reason is insufficiently critical (Poorthuis, 2015), and reductionism is hailed and applauded. Individual knowledge and knowledge accumulation have become more important than understanding each other. Not knowing, or not yet knowing, leaves a person vulnerable – and this also applies to *non-mainstream thinking*.

Consequently, many universities remain lacking when it comes to laying the foundations for a *'caring society'*, even though many managers claim that they are in fact securing such foundations. Indeed, some time ago a large billboard was erected in front of the VU's main building with a slogan stating just that. Others show a better sense of reality and avoid the issue by stating, rather shamelessly, that things could also be a lot worse – as can be seen elsewhere.

The initiative to design the Dutch National Research Agenda via the process of *societal sensing*[2] is a commendable one. Members of society are invited to share their thoughts and ideas about socially relevant research questions. But what topics will ultimately be included in the final agenda? And how can we make sure to find solutions that will prove to be more or less effective and appropriate rather than answers that would seem be correct in theory only? We need to find solutions that carry meaning rather than solutions that are expedient or, even worse, that have *added value*. We need meaningful solutions that concern living together, *to improve the human condition* (Arendt, 1958). The situation is very aptly illustrated by the following call to action in which individuals and organisations are invited to help compile the National Research Agenda:

On 31 March next, the National Research Agenda website will be launched. For a period of four weeks, individuals and consortia of scientists, commercial businesses and/or social organisations will be offered a chance to submit their

2 *Societal* is the pedantic alternative to *social*. They both mean 'pertaining to society,' but as the latter word, first attested in the Middle Ages, was increasingly used in the modern era to refer to interpersonal contact rather than in the context of complex forces within human populations, *societal* appeared in the latter part of the nineteenth century as a more serious, scholarly alternative. It is mostly seen in such usage and is otherwise considered pretentious.

proposed research questions. The initial selection of questions will take place on the basis of the following questions:

- *Can the question be answered through scientific investigation?*
- *Is the answer to the question currently lacking?*
- *Can the question be linked with research expertise or a specific research focus in the Netherlands?*

What is important here: answers or solutions? Do we have special PIN numbers for social issues? Are we concerned with research strength and expertise, or are we concerned with making society stronger? There is a clear need for universities to become more democratic and more internally *connected*, and they will have to find ways to fulfil the potential of *our* scientific endeavours in society. This will prove to be the most effective form of *governance* imaginable for society. The current essay explores how this might work, without pretending that what it presents would be the one and only way. Why do I do this? I do this because I consider it my moral duty to fight institutions and convictions that pose a threat to human dignity and to make a meaningful contribution to the academic debate concerning the *teaching-research nexus* (cf. Boyer, 1990; Ramsden & Moses, 1992; Hughes, 2005)[3]. I know that I cannot execute the first task completely on my own, but I do know that I can rely on the following (from Joseph Jaworski's *Synchronicity – the Inner Path of Leadership*):

The Weight of Nothing

'Tell me the weight of a snowflake,' a coal-mouse asked a wild dove. 'Nothing more than nothing,' was the answer. 'In that case, I must tell you a marvelous story,' the coal-mouse said.

'I sat on the branch of a fir, close to its trunk, when it began to snow – not heavily, not in a raging blizzard – no, just like in a dream, without a wound and without any violence. Since I did not have anything better to do, I counted the snowflakes settling on the twigs and needles in my branch. Their number was exactly 3,714,952.

3 See http://trnexus.edu.au/index.php?page=bibliography for an excellent overview of academic publications on the teaching-research nexus.

When the 3,714,953rd dropped onto the branch, nothing more than nothing as you would say – the branch broke off.

Having said that, the coal-mouse flew away.

The dove, since Noah's time an authority on the matter, thought about the story for a while, and finally said to herself, 'Perhaps there is only one person's voice lacking for peace in the world.'[4]

Window of opportunity

A future-proof university is deeply committed to truthfulness, dignity, offering service and addressing ignorance as complementary core themes of *experiential learning*. Giving people a chance to be nourished by these themes supports and enriches the process of truth seeking. It will enhance the development and formation of teaching staff as well as many future generations of students. This deliberate choice calls for a university environment in which the gap between research and education disappears, in which research can only be *temporarily disconnected* and where justification is found in existential rather than fundamental perspectives. Research does not need to be useful *per se*, but necessary and meaningful.

A university that is ready for the future places higher demands on incoming students and their progress, replaces lectures with interactive classes, tutorials or seminars and has lecturers who operate as '*more capable peers*'. Rather than act as socially dominant figures or teachers who dictate thoughts, they are the ones to create the necessary conditions for true learning. This is the only form of knowledge management that, from an instrumental perspective, can be termed valid as well as ethically responsible. Teachers acting as *more capable peers* allow their students to waver, but they will never allow them to stumble and fall.

At future-proof universities, psychological contracts based on mutual respect and attention (Roeland, 2015) will no longer require the use of deadlines, and students will be given the opportunity to demonstrate their skills in ways that are appropriate and in line

4 Jaworski, J. (2011). *Synchronicity – the Inner Path of Leadership* (second expanded edition), p.196. San Francisco: Berrett Koehler Publishers.

with their role of *junior researcher* and *junior practitioner*[5]. A permanent point of focus for instructors is the need to ensure sufficient variation with respect to the way in which instruction is offered. At first sight, *unhappy learning* would seem to be a contradiction in terms, but this is absolutely not the case.

The ease with which people can ask for help *and* receive it forms a future-proof university's framework to ensure rich learning processes (Grodal *et al.*, 2015) in which teachers and students, masters and apprentices, interact in fruitful exchanges. A university that is prepared for the future offers students a safe haven: a place where they can learn to rely on themselves and develop independent thoughts instead of simply following traditional ideas. Such a university is a house as well as a home; it is a place where limiting convictions will erode. As the form in which content is offered has a significantly greater impact on the development of students than content alone, we may conclude that it is predominantly form that determines formation. Precisely this is what we call *Bildung* (Groot, 2015). Students' increased confidence and determination will prove to be the most important precursors and foundation stones of a society in which trust replaces distrust. In this way, the academic community will develop from a *community of learners* into an ecosystem of pioneers who pave the way. Dialogue and deceleration rather than widespread debate will lead to true meaning.

A university that is built for the future operates along lines based on knowledge and insight, ability, empathy and mastery, as illustrated in Table 1 below. Knowledge and insight give us a clearer picture or a better understanding of the ways in which things have materialised or may materialise. This does not, however, always ensure a larger action repertoire, and some of this knowledge is fleeting. Extending action repertoires requires engaging in practice situations in which new knowledge can be used. Practice allows individuals to experience usefulness and meaning, not solely for themselves but also for others, thus making them consciously competent (or consciously *in*competent, as the case may be). Once usefulness or pleasure has been perceived and experienced, it becomes possible to engage in new situations, cases when this 'single'

5 I expect these two roles to show increasing overlap as time progresses, because more and more tasks in our modern world are related to *search* work. Today's researchers are required to demonstrate an ever-larger range of competencies, and twenty-first century students will have to hone their skills related to '*search strategies in context*'.

knowledge is triggered anew, with a sense of trust and confidence. One thus develops competencies. The ability to mobilise 'complex' knowledge or skills in situations that require their application calls for mastery and a primary instinct to determine the nature of the situation (*sensing*). In fact, this concerns an integral and instantaneous understanding of the *boundary conditions* associated with knowledge and with insight into available action repertoires, including the social conditions needed to translate this knowledge into productive action.

If the process of sensing is perfectly executed, this will create a *flow* followed by the desire to return to the state of mastery. *Flow* is not useful: it is pleasurable. One of my mentors taught me that good instructors always prepare extremely well, but also that events and developments in actual practice will always unfold in the way that they *should* unfold, irrespective of preparations. Good instructors 'sense' the situation and engage with their audience; they are certainly no slide pushers or PowerPoint slaves. They heavily rely on their intuition, which will likely have been built, either in part or fully, upon insights gained by experience. What lies at the heart of the trust that is generated in this way is intensive practice. That said, we should also bear in mind that good quality does not lie in numbers or amounts, but that there is a lot of good to be had with quality. This has its consequences for future-proof universities, where cognitions are not superficially dealt with in their droves, but carefully considered during intensive practice and in modest numbers based on the nature of these cognitions, for instance related to the formulation of a particular problem and the subsequent definition of effective (re)search strategies in a number of highly varied contexts.

Productive working relationships between students and instructors allow a university that is well prepared for the future to develop vital networks that are extremely meaningful not only for society as a whole but also for individual sectors to which these networks have spread. Such sectors may include Amsterdam's commercial Zuidas population, but also more romantic cloud hunters and idealists (Schoenmakers, 2015). Once it has become firmly rooted in society, such an ecosystem can offer exactly the right type of nourishment that is needed to supply a future-proof university with new questions and topics for investigation. It is a vital condition for '*earth stewardship*' (Dickinson *et al.*, 2012), it contributes to the deconstruction of many ivory towers flying the flag of 'fundamental research' with an exclusive and isolated focus, and it enhances the accumulation of funds via contract research. The academic ideal of a university that is truly free from church *and* state influence has come nearer than ever before. At the same time, however, attention for church and state has become more extensive than ever before. Institutions and convictions are scrutinised, valued and connected. Polarisation is eliminated, and processes that turn society into a meritocracy are halted, although this is certainly no plea for the introduction of a system in which backgrounds determine social status and position: far from it.

Function	*Development*	*Result*
Knowing	Cognition	Imagination
Ability	Competence	Confidence
Control	Mastery	Aspiration

Table 1. From Cognition to Mastery.

Truthfulness

The French philosopher Michel Foucault spent a large number of his classes on *parrèsia*, the virtue of speaking candidly and boldly. He introduced the concept as follows: 'those who use *parrèsia*, the *parrèsiastès*, are individuals who voice all of their thoughts: they hide nothing, but they open up to others, heart and soul, in order to speak their mind'. What is concerned here, according to Foucault, is the courage to air one's convictions and to accept any possible associated risks (Trouw, 12 March 2015). A future-proof university enables students and instructors hailing from all of its faculties and departments to participate in classes where everyone can see and hear what the *parrèsiastès* have to offer them. We thank the opportunity to enjoy this interfaculty manna for the soul to our Humanities community, making their discipline a true Major Study Area again. Humanities scholars were also the first to found the predecessors of today's universities, although these were not quite shaped in the way they are today.

Towards the end of the twelfth century, Europe saw the rise of its first universities. This meant the establishment of educational guilds that operated alongside monasteries to form what is known as the *universitas*. At the time, 'guild' was a generally used term to refer to groups of people who worked together to achieve a certain goal or to safeguard common interests. The *universitas* initially operated as a guild, although its scholars were long regarded as representatives of the clergy. The *universitas* aimed to protect its members and to secure educational interests in general: nothing could be nobler than that.

A university that is future proof continues to expand its horizons and look further, but it also trains its students to look back and reconstruct the past (Groot, 2015); it makes them aware of heritage, good stewardship and *path dependencies*, the consequences of past and future decisions. What is important is the creation of a *sense of heritage* (cf. Grever, 2016), a sound feeling for history and culture. Important roles to be played are those reserved for historians, anthropologists, biologists and economists.

At future-proof universities, it is not the production of scientific output that is given primary importance, but contributions towards the further good of society: contributions with *societal impact*. This is no mere sales advantage or focus point that is easily traded for something else, but an ongoing theme that is wholeheartedly embraced by all. Staff members literally live it, every day of their professional careers. They take the time they need to share their thoughts in carefully considered ways with truthfulness, candour and boldness, in speech or in writing and in forms which they deem to be appropriate. As Alexandre Dumas preferred to describe it: 'Time is a friend and a fellow traveller; haste is a poor counsellor'.

At universities that are ready for the future, debates and expositions are just as important as papers listing the results of empirical study. In this way, freethinking and -writing is revitalised. The academic population is not primarily concerned with monodisciplinary attempts to add yet another text to a mountain of publications, but first and foremost with multidisciplinary contributions for defining and addressing issues that match the identity which the university has built for itself and which forms its anchor points in the pursuit of progress. Like poets, we have to wait for beauty, and therefore truth: 'Whatever the imagination seizes as beauty must be truth – whether it existed before or not' (Keats, 1820). We should cease the hunt for theories; any of its unexpected additional fruits is hailed as welcome by-catch. There is no need to motivate experts to deliver expert work: it's what they want to do most of all, and they are never truly satisfied. A university that is future proof shares a collective conviction that returns on investments require 'assistance to make progress'. Nothing is more rewarding than making progress (Amabile & Kramer, 2011), except the ability to assist in this process.

At universities that are ready for the future, the accreditation and re-accreditation of programmes is a formality. In all respects, their curricula are firmly grounded and the subject of ongoing scrutiny, after which they may be recalibrated based on experiences and changes in their environment. This is what I call obtaining a greater '*sense of purpose*', a true understanding of goals and destinations. This can be achieved via a '*sense of synchronicity*', attention for current affairs and developments as they unfold paired with a focus on the various opportunities this offers. The sum total of a programme's elements and distinguishing features, from the smallest detail to the overall picture, forms its test matrix. This means that comprehensive and hefty accreditation reports will become redundant – and will disappear, as will the accompanying box-ticking activities that leave people with a false impression of being *in control*. And then there is this other notion, termed ´alignment': an unpleasant word that often proves to be completely meaningless. 'Control' is an illusion; 'commitment' is not. Initiatives are not formulated on paper, but they are taken in universities with a future. Professors briefly introduce themselves, with a particular emphasis on *why* they are there. Mature *purpose*, shared commitment and practice will weld the parts of the structure firmly together in order to form a service-minded, coherent and organic whole. We may well ask ourselves if this idea would be somewhat naïve, but the answer would be 'no, it would not'; it offers us a wonderful and inviting perspective. A particular education programme's specific elements, great and small, partly determine the research agendas for practice groups in which there is deep mutual respect for craftsmanship. Departments are not officially formalised or strictly delineated, but they transcend the boundaries of their own university – and, as a matter of fact, of *the* University. Practice and reality offer more science than anything else in the world. Craftsmanship cannot be organised; it organises itself.

A future-proof university's PhD students are students who keep a keen eye on the outside world. They have a clear view of what goes on in this world and they operate from the outside inwards. They are fuelled by issues that touch them and that invite them to take the first step on their research journey. They write whatever they want to write, and they learn to listen to themselves. The interaction between tacit and explicit knowledge will lead to experiences of unthought knowing. Not only do they stand on the shoulders of others, but they also stand shoulder to shoulder with others.

A future-proof university's professors are appointed on the basis of eye witness testimonies. These testimonies should offer a convincing picture of the ways in which the master influenced societal developments, visibly or invisibly, along lines laid down in the four core themes mentioned earlier and on the basis of this person's mastery and craftsmanship. The truthfulness of these testimonies is the ultimate acid test. Permission to occupy the post of professor will require more than is presently the case, and those who are entrusted with the tasks involved in the profession will take the initiative in emphasising the power of *witness-thinking* rather than *aboutness-thinking* (Shotter, 2005). They will set out to halt developments where researchers distance themselves from empiricism (Reason, 1998) while still airing all kinds of statements founded on practice-based causal relationships that have very little instrumental validity.

Universities that are ready for the future make important new steps forward without 'rubber stamping' efforts and developments. Career paths are considerably simplified, and the system of league tables distinguishing *lecturer-1, lecturer-2, lecturer-3 and lecturer-4* positions is abandoned. This system is one of the outcomes of the process of alienation that we are witnessing today. Our current job classification system, a product stemming from the equally alienated discipline of Human Resource Management (aptly illustrated by the dubious concept of *e*-HRM), has made notable contributions towards deconstructing our academic dwelling place. It forms an outright threat to liberté, égalité and fraternité or its often related – and perhaps even more appropriate – partners *amitié, charité* and *unité*. It is this unity that protects the simplicity we seek to safeguard.

True and truthful managers monitor and guard unity. At universities that are fit for the future, they keep a clear eye on the following three indicators: i) the quality of incoming students, a process for which they, too, carry responsibility, ii) staff well-being (cf. Boselie, 2014) and iii) freedom not only from state and church influence, but also from egos and megalomania. These three areas in particular form the focus of their attention and attempts to ensure progress. This can be achieved by prioritising a sense of serving, namely by serving their staff.

Serving employees first and protecting purpose above all! No person is mightier than the parts that make up the whole. At future-proof universities, this is no idle talk, but it means action: 'theory in use' rather than 'espoused theory' (Argyris & Schön, 1974). True managers 'put their money where their mouth is'. They do not rigidly respond to black swan events, and they do not blindly follow hypes and trends; they are steadfast and unwavering in their mission. Their destination, their *sense of self*, is to learn how to serve others by serving them and society through them. Creating *Sense of Serving*: wouldn't this be a wonderful alternative to *´looking further*', the VU's slogan in use? After all, it is people who form the main impact factor.

Sense of Serving and Ignorance

In his book 'Helping', professor emeritus and sociologist Edgar Schein describes what he calls the *helping paradox*: the wish to help others may be strong in many, but in practice requests for help from others often remain remarkably limited, at least initially. This is where the paradox is found. Schein demonstrates that most helpers are not helpful towards others, but instead mainly driven (by various different motives) to satisfy their own need to provide help. According to Schein, helping involves offering someone else – with this person's consent – an opportunity to do what this person cannot do on his or her own. What Schein elaborates as *humble inquiry*, careful observation and listening followed by an open-minded approach towards determining the type of help required, is what helpers need to offer the right assistance in a convincing way. It partly shapes and determines the knowledge object that I call Sense of Serving. Mastering the technique of the *humble inquiry* is an exercise in accessing one's ignorance, in diligent inactivity and indolence (Saggurthi & Thakur, 2013). This is a methodological lesson that everyone, not only social scientists, should take to heart in order to enhance their own formation. At universities that are ready for the future, coming to grips with *humble inquiry* will function as a stumbling block for all of their students; for instructors, it will be a recurring theme in the University Teaching Qualification programmes BKO and SKO.

Universities that are future proof do not focus on reality as if it were a single phenomenon, but they study its plurality and the many ways in which reality manifests itself. Reductionists will be left scratching their heads. In this way, monodisciplinary and self-absorbed behaviour can be prevented. Moreover, younger as well as older generations will be able to discover the developments with which they feel connected the most and to which they would like to contribute. In sum, hearts and minds will be given a solid position; in addition, it cannot be denied that 'being roughly right is better than being precisely wrong'. This means that a university that is ready for the future is also a campus location of the *School of Life*.

In the School of Life, students learn that being meaningful to others is what connects and unites people – and that this makes society strong and resilient (societal resilience). It is fulfilling, and this fulfilment offers something invaluable: love and gratitude. Table 2 includes the 'superlative' form of control and mastery: meaning – finding your destination. The development process concerned here is best described as identification: people do what they like doing best and what suits them best. It is something they feel they must do, because it is their destination. Such processes are often triggered by certain *focal events*, and they frequently stem from action-inducing sources such as crises or feelings of displeasure. In this way, the urge to secure particular positions is suppressed, at least for the time being.

Dignity

Need I say more? The Charter of the United Nations (1945) and the Charter of Fundamental Rights of the European Union (2000) mention human dignity as a key term, even though they do not specify it. This could be considered somewhat sloppy, but it could also be

Function	*Development*	*Result*
Knowing	Cognition	Imagination
Ability	Competence	Confidence
Control	Mastery	Aspiration
Meaning	Identification	Destination

Table 2. From Knowing to Meaning, Identification and Destination.

argued that what is referred to here is in fact a *sensitising* concept that needs continued calibration in view of its ever-changing context and the current state of affairs. It must never become a *definitive* concept. In fact, it offers an extraordinary opportunity for Law School staff to plead the case for future students in future-proof universities to be trained on the topic of human rights and to learn how act in defence of humanity. It also offers a unique opportunity for linguists to show students how demeaning and degrading situations can be 'framed' and removed out of sight. In addition, it will help students increase their language awareness, so that language may well and truly become their kind of thing – a welcome development in today's world of small messages. Political scientists will voice the ideas of engaged and committed contemporaries such as Martha Nussbaum, who never tires of showing that a just society cannot function without love.

At universities that are ready for the future, students are clearly seen and acknowledged: '*sense of belonging*' opens up everyone. Students are made aware of their mind-sets and assisted in developing productive forms of action to address their doubts, including moral or political doubts. Assistance will decrease as skills will improve over time. We also propose deep respect for identity and diversity. This means putting a stop to uniform attempts to transfer knowledge that has been codified with maximum efficiency to large groups of students in large lecture halls in which these students are introduced to their instructor's supposed knowing and in which there is no room to offer the safe environment, or rather containment, that is required for rich and rewarding interaction. That being said, let us reconsider the situation of students being 'introduced' to knowledge; knowledge 'eloquently delivered' is no guarantee for 'a lot of new things learned'. Many of today's student attend classes for social reasons or to give themselves the idea of being involved with their programme's curriculum. Learning outcomes are low, and before you know it you get your knuckles rapped, either by your fellow students or, if you are a lecturer yourself, from your audience when its members feel that you are not sufficiently entertaining or that your class lacks structure. Now, wasn't this something that students had to learn? At universities that are fit for the future, learning outcomes are high, partly thanks to attempts to enhance on-the-job-learning, particularly in the master stage. This means, for instance, that students are given the opportunity to contextualise their own learning processes in view of the position they wish to adopt in the labour market. This does not usually concern PhD tracks, but the option must not be ruled out during the early stages of this process. It also means that we return to students the level of control they deserve to have over their own learning process.

Reconciliation of education and research

'Riding the waves of culture' is a best-selling book written by Vrije Universiteit Amsterdam professor Fons Trompenaars, co-founder of SERVUS, the VU's Centre for Service and Leadership. Using 'servus', one greets others in a much friendlier tone – and with greater engagement – than is the case with the more formal use of 'good morning', 'good afternoon' or 'good evening'. 'Servus' is used for saying hello as well as goodbye. At future-proof universities, 'servus' signifies a standard form of behaviour; it carries a *sense of serving* that makes students in all departments and faculties aware of their possible impact across cultures. It offers a form of practice to reconcile dilemmas and to develop a *sense of place*, and students will experience that dilemmas are mostly, and perhaps even always, false trade-offs and that they are in fact paradoxes. To illustrate the point, dilemmas often involve choice: one may prefer one option to another (or-or), but it often turns out that through the resolution of one dilemma something else can be realised, too (through-through). Considered more creatively, dilemmas are paradoxes; therefore, *through-through* rather than *or-or* notions will resonate with students. The ability to reconcile dilemmas has founded my conviction that 'caring about our students' not only has excellent role model effects, but also forms a major component of the way in which education and research are reconciled. We can improve research through improved education – ánd vice versa, of course, but the latter is a generally accepted fact.

Finally, but not quite lastly, what lies at the very heart of a future-proof university is its teacher training programme: preparing students to become secondary-school teachers. This programme will no longer be banned to the academic periphery. At the same time, each individual faculty will accept the responsibility for training teachers operating in secondary education. Teacher training programmes will become a specialised master's programme rather than a separate or a post-master programme. In addition, trainees from various disciplines will be given the opportunity to establish effective collaboration in order to ensure continued progress.

Lastly, the combination 'you are an excellent teacher, but …', something that is frequently heard during performance reviews nowadays, discussions and evaluation meetings, has become a thing of the past at future-proof universities as well as in the outside world. The fact that this type of feedback is a regular phenomenon is downright disgraceful and unbelievable. '*Great teachers care*': this is all we need, and it is all that society needs. *Caring* is by no means soft and sentimental; it can be rock solid, it is meaningful, and it is of existential importance.

A lack of funds?

Staff-student ratios show no signs of improvement, but the idea that insufficient funds are available for small-scale academic education is incorrect. What matters first and foremost is the quality of the interaction between masters and apprentices, mentors and students. Much can be achieved with small pushes. In fact, small pushes are usually all that is needed. New technologies can be of help, but they should not be used as a budgetary escape mechanism. IT may establish links between students and instructors, but it does not truly connect them. The physical presence of mentors and fellow students will remain a vital feature of universities that are future proof. We shall have to reconsider the allocation of resources. We should not continue 'scrapping' these, but we must think about 'replacing' them. We need to plant seeds in fertile soil. In Figure 1, I have tried to illustrate this point by visualising the flower that will grow from these seeds. It has a thin stem symbolising individual and collective knowing, and although this stem may be delicate, it acts as a strong support for highly promising developments. It is not *knowing* that will further develop society and improve conditions for living together, but *sensing*. For me, this embodies the modern version of Alexander von Humboldt's ideal of Bildung.

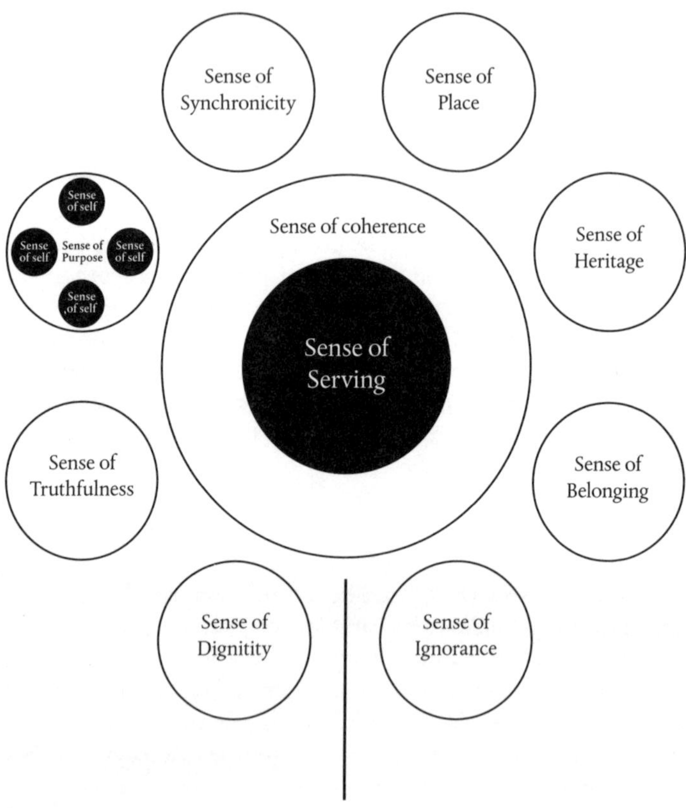

Figure 1. Sensing.

Academia has not yet sufficiently moved to the stage of transition that is called for today. By merely paying lip service to the necessary and fundamental review of the role of universities in modern society, we are making insufficient progress. What is needed is a velvet revolution. This type of revolution will be launched from within the academic world itself, not to be broken by anyone. Love and devotion cannot be broken. Instead of balancing education and research, we need to reconcile the two (cf. Wals *et al.*, 2014). The same holds true for Education, Culture and Science. This is our core task.

Of course, we should also create positive yields. Incidentally, the development of this word in English comes via Anglo-Saxon 'gildan' ('to pay', 'to give', 'to restore'); today, the word has come to mean 'to give in return'. In this context, some are speaking of creating what may be termed Gross Academic Value (Winter, 2015): a brave initiative, but one that also exposes us to the risk of launching a new collective hunt for ways to measure this value. Before we realise it, we may end up in structures that are similar to those we are already familiar with: modernised and completely up-to-date versions of a well-established production plant as opposed to a vivid community of learners. My preference would be 'meaning', or perhaps 'meaningful knowledge', to show that academically self-absorbed behaviour should be monitored critically and that the university departments that have developed into sheltered workshops for the highly educated should be closed or reformed.

Universities that are fit for the future introduce reforms and adopt some of the inspirational features that typify the University of Humanistic Studies. These universities are cherished by many, particularly by their alumni, and they measure their performance on the basis of their ability to avoid publicity. Modesty makes them beloved.

Future-proof universities have no *hall of fame*; they have a *hall of humility*, a *yad vashem*. This chapel of silence promotes humility, and for me it illustrates the resounding beauty in the words of the prophet Isaiah (56:5): 'I will give them an everlasting name that will not be cut off'. We must never forget, and we must forget no-one. It is not the top of the academic pyramid that we should seek, but a sense of service. This is what stems from the source of everything: love.

At future-proof universities, teaching continues to be the foundation stone, and experts from all over the world will be offered a chance to share their deep knowledge, for instance via Massive Open Online Courses (MOOCS). Other people's expertise will be redirected towards creating conditions for development and for making sure that students literally experience how the vast potential of knowledge can be realised, in part also by training interpersonal skills (Bedwell *et al.*, 2014). This requires a greater *Sense of Serving* on the part of instructors. The message may be considered an SOS message, but at the same time it is one that carries immense personal meaning. Teachers will experience the phenomenon of double loop learning processes, where personal learning defines and shapes the realm of student learning. Shared meaning is what connects these learning loops. *Qui docet discit*: he who teaches learns.

A university that is future proof is not only receptive of the potential of new generations and capable of distinguishing trends and undertows in these groups, but it is also naturally inclined to respect what has to remain unchanged. It remains in constant connection with its roots and thus honours the Universitas. It shapes today's *homines universales* via liberal education (Zakaria, 2015), and it returns to them – our students – the control they deserve to have over their learning processes.

This contemporary university has music that heals; it also promotes the cardinal virtues. It shines a different light on society whenever this is needed. It is an *alma mater*, led by servants of society who personify faith, hope and charity. These servants are already in our midst. Let us cherish them. This is my small message. We also have a meaningful slogan: 'foundations for a caring society'. This is a slogan we must never replace. It only takes a few little pushes to open up a window of opportunity for universities to become liberated and free, something that is of vital importance for society as a whole. These little pushes will spark a process of autopoiesis, of self-organisation, with Sense of Serving as a beacon guiding the academic ship towards the great and wonderful unknown.

Via *sensing*, a future-proof university serves the purpose of realising the potential of *knowing*. It can put itself into perspective while helping people fulfil their potential by caring for them. It promotes societal wellbeing by being together in a truthful community. It lays the foundations for social existentialism.

Universities with promising potential for the future will rid themselves of all the systems and symbols associated with the obsessive hunt for academic top positions. They will halt today's collective rush for gold in Academia, knowing that craftsmanship is capable of organising itself. They literally live and breathe *Sense of Serving*, which guides and acts: it connects, supports, transforms and adds value.

Let us make a start by liberating ourselves from everything that threatens our academic freedom 'to improve the human condition' (Arendt, 1958 and 2013). Let us collectively scrap the idea that we should 'publish or perish'. This can be done as of today by placing the development of our students firmly at the centre of our ambitions, in our words as well as in our actions. This would secure their modern *Bildung* without any crippling preconceptions or predefinitions and without developing less than solid curricula or introducing a culture designed to let a hundred flowers bloom.

There is one particular opportunity that is waiting to be seized, namely the opportunity to re-adopt a pioneering role in Academia and to present ourselves, justly and with a modest sense of pride, as the type of university that lives up to its name of *Vrije* Universiteit: a Free University laying foundations for a caring society that are solid as a rock. It only requires a sense of serving.

Post Scriptum

In my academic career, I have always felt a primary responsibility for the formation and support of my students. Despite all of my teaching awards and the positive feedback that I have received from my students over the years, I still feel a healthy tension when I prepare for working with them. For me, this is the ultimate proof of the responsibility that I feel and the complexity of the profession. At the same time, however, there have been increasingly loud calls over the past decade to 'increase the number of publications and to make greater contributions to theory' without any form of societal justification. We cannot close our eyes to this development. Things must change and become more tuned to societal needs.

References

Adler, P. (2016). 2015 Presidential Address: Our Teaching Mission. *Academy of Management Review, 41* (2), 185-195.

Amabile, T., & Kramer, S. (2011). *The progress principle: Using small wins to ignite joy, engagement, and creativity at work.* Harvard Business Press.

Arendt, H. (1958). *The human condition.* Chicago, University of Chicago Press.

Argyris, C., & Schon, D. (1974). *Theory in practice: Increasing professional effectiveness.* San Francisco: Jossey Bass.

Baars, L. van (2015). Is de student tevreden? Dan doe je iets verkeerd. Interview with Gert Biesta in *Trouw,* 9 April 2015.

Bedwell, W.L., Fiore S.M., & Salas, E. (2014). Developing the Future Workforce: An Approach for Integrating Interpersonal Skills Into the MBA Classroom. *Academy of Management Learning & Education,* 13(2), 171-186.

Boselie, P. (2014). *Strategic Human Resource Management: A balanced approach.* Tata McGra Hill Education.

Boyer, E. (1990). *Scholarship Reconsidered: Priorities of the professoriate.* New Jersey: The Carnegie Foundation for the Advancement of Teaching.

Dickinson, J.L., Shirk, J., Bonter, D., Bonney, R., Crain, R.L., Martin, J., & Purcell, K. (2012). The current state of citizen science as a tool for ecological research and public engagement. *Frontiers in Ecology and the Environment,* 10(6), 291-297.

Flikkema, M.J. (2014). Geef docenten perspectief. See: http://www.advalvas.vu.nl/opinie/geef docenten perspectief.

Grever, M. (2016). Zonder geschiedenis varen we stuurloos door de wereld. *Trouw* 11/3, Opinie, p.21.

Grodal, S., Nelson, A.J., & Siino, R.M. (2015). Help-seeking and help-giving as an organizational routine: continual engagement in innovative work. *Academy of ManagementJournal,* 58(1), 136-168.

Groot, G. (2015). Gun mij m'n Bildung. *Trouw* Letter & Geest, 16-19.

Hattie, J., & Marsh, H.W. (1996). The relationship between research and teaching: A meta analysis. *Review of Educational Research,* 66(4), 507-542.

Hughes, M. (2005). The mythology of research and teaching relationships in universities. In R. Barnett (Ed.), *Reshaping the university: New relationships between research, scholarship and teaching* (pp. 14-26). Buckingham: Society for Research into Higher Education/Open University Press.

Huijer, M. (2015). Bildung, de tandeloze tijger – tegen het rendementsdenken in het onderwijs. *Trouw* Letter & Geest, 4-7.

Jaworski, J. (2011). *Synchronicity – the Inner Path of Leadership* (second expanded edition), p. 196. San Francisco: Berrett Koehler Publishers

Kleijn, J. de (2014). Wetenschapsvisie toont gebrek aan visie. *Trouw,* 9 December 2014.

Kuijper, L.D. (2015). *Het onmeetbare van idealen is geen reden om het meetbare tot ideaal te verheffen.* In: Visies op het universitaire onderwijs – bijdragen van de genomineerde docenten voor de Van der Duyn Schouten Onderwijsprijs. Vrije Universiteit, September 2015, p. 27-30.

Maex, K. (2015). De grootste uitdaging. THEMA hoger onderwijs, 4, p. 44.

Poorthuis, M. (2015) Benader ook de rede kritisch. *Trouw,* 26 March 2015.

Roeland, J. (2015). *Met aandacht doceren.* In: Visies op het universitaire onderwijs

– bijdragen van de genomineerde docenten voor de Van der Duyn Schouten Onderwijsprijs. Vrije Universiteit, September 2015, p. 33-35.

Saggurthi, S., & Thakur, M. (2014). Usefulness of Uselessness: a case for negative capability in management. *Academy of Management Learning & Education*, amle 2013.

Schouw, G. (2015). Integriteit kun je niet meten, wel bespreken. *Trouw*, 3 April 2015.

Schein, E.H. (2013). *Humble inquiry: The gentle art of asking instead of telling*. Berrett-Koehler Publishers.

Schoenmakers, M.M. (2015). *De Wolkenridder*. Uitgeverij de Bezige Bij.

Verdonk, P. en Abma, T. (2015). Herbezinning nodig op rol universiteit. *Trouw*, 6 March 2015.

Verhulst, D. (2015). *De zomer hou je ook niet tegen*. National Book Week Gift 2015, Uitgeverij Atlas Contact.

Wals, A.E., Brody, M., Dillon, J., & Stevenson, R.B. (2014). Convergence between science and environmental education. *Science, 344*(6184), 583-584.

Winter, J. (2015). Rede ter gelegenheid van de opening van het academisch jaar 2015-2016.

Zakaria, F. (2015). *In Defense of a Liberal Education*. Ww Norton&Co.

About the Editor

Dr. Meindert J. Flikkema (1971) obtained his degree in Econometrics from the Rijksuniversiteit Groningen. He continued his career at Vrije Universiteit Amsterdam where he published a dissertation on innovation in the service sector. He was awarded the Faculty of Economics and Business Administration's Education Prize in 2005, 2014 and 2015, and the Van der Duijn Schouten Education Prize from Vrije Universiteit Amsterdam in 2015. In 2016 he was among the five nominees of the 'national teacher of the year award' according to the Interstedelijk Studenten Overleg (ISO), after a pre-selection of twenty candidates. Meindert Flikkema currently works for Vrije Universiteit Amsterdam as associate professor, and he is the academic director of the *Amsterdam Centre for Management Consulting*.